S0-ANN-477

The Color of Opportunity

Haya Stier and Marta Tienda

The Color of Opportunity

Pathways to Family, Welfare, and Work

THE UNIVERSITY OF CHICAGO PRESS
CHICAGO AND LONDON

Haya Stier is senior lecturer in the Department of Labor Studies and the Department of Sociology at Tel Aviv University. Marta Tienda is the Maurice P. During '22 Professor of Demographic Studies, professor of sociology and public affairs, and director, Office of Population and Population Research at Princeton University.

The University of Chicago Press, Chicago 60637
The University of Chicago Press, Ltd., London

Printed in the United States of America
10 09 08 07 06 05 04 03 02 01 5 4 3 2 1

ISBN (cloth): 0-226-77420-1

Library of Congress Cataloging-in-Publication Data

Shṭayer, Ḥayah.
The color of opportunity : pathways to family, welfare, and work / Haya Stier and Marta Tienda.
p. cm.
Includes bibliographical references and index.
ISBN 0-226-77420-1 (alk. paper)
1. Poor—Illinois—Chicago. 2. Minorities—Illinois—Chicago—Social conditions. 3. Chicago (Ill.)—Social conditions. 4. Chicago (Ill.)—Economic conditions. 5. Family—Illinois—Chicago. 6. Poverty—Psychological aspects. I. Tienda, Marta. II. Title.
HV4046.C36 S5 2001
305.569'09773'11—dc21 00-009125

♾ The paper used in this publication meets the minimum requirements of the American National Standard for Information Sciences—Permanence of Paper for Printed Library Materials, ANSI Z39.48-1992.

To the men in our lives . . .

Rami, Asaf, Omer—H.S.
Luis, Carlos—M.T.

Contents

	List of Illustrations	ix
	Preface	xvii
	Acknowledgments	xxi
One	Poor People, Poor Places	1
Two	Chicago: Economic and Social Transformation of an Urban Metropolis	28
Three	The Study Population	71
Four	Family Matters: Turning Points from Orientation to Procreation	92
Five	Doles and Safety Nets: Public Assistance and Income Support	130
Six	Makin' a Living: Employment Opportunity in the Inner City	165
Seven	The Contours of Opportunity	218
Appendix A	Design of the Urban Poverty and Family Life Survey	239
Appendix B	Methodological Appendix	245
	References	275
	Index	283

Illustrations

Tables

Table 2.1 Racial and Ethnic Composition of Metro Area: Cook County and Six-County SMSA, 1970–90 34
Table 2.2 Educational Attainment by Race and Hispanic Origin: Cook County Residents Ages 25+, 1970–90 41
Table 2.3 Labor Force Status by Race, Hispanic Origin, and Age: Cook County Men Ages 18–64, 1970–90 43
Table 2.4 Labor Force Status by Race, Hispanic Origin, and Age: Cook County Women Ages 18–64, 1970–90 45
Table 2.5 Poverty and Program Participation by Race and Hispanic Origin: Cook County Families, 1970–90 46
Table 2.6 Changes in UI-Covered Manufacturing and Nonmanufacturing Employment: Chicago Metro Area, 1970–90 49
Table 2.7 Descriptive Statistics by Neighborhood Type 57
Table 2.8 Neighborhood Mobility through Ecological Categories 64

Table 2.9	Descriptive Statistics by Pattern of Neighborhood Change	66
Table 3.1	Distribution of UPFLS and NSFH Respondents by Ethnicity and Sex	82
Table 3.2	Sample Characteristics by Ethnicity and Sex: Urban Poverty and Family Life Survey	84
Table 3.3	Sample Characteristics by Ethnicity and Sex: Urban Residents, National Survey of Families and Households	88
Table 3.4	Sample Characteristics by Ethnicity and Sex: Low-Income Urban Sample, National Survey of Families and Households	89
Table 4.1	Selected Characteristics of Families of Orientation: Parents Residing in Chicago's Inner City and U.S. Cities	96
Table 4.2	Ethnic and Sex Variation in Parental Supervision during Adolescence	100
Table 4.3	Selected Indicators of Problem Behavior during Adolescence	103
Table 5.1	Current and Recent Welfare Participation Rates: Parents Residing in Chicago's Inner City and Urban United States	134
Table 5.2	Welfare Participation Rates by Neighborhood Poverty Rate: Women and Men in Chicago's Inner City	137
Table 5.3	Preferences for Welfare versus Work by Use Status: Inner-City Parents by Race and Ethnicity	158
Table 5.4	Attitudes toward and Knowledge about Welfare by Use Status: Inner-City Parents by Race and Ethnicity	159
Table 6.1	Employment Status Distribution of Men and Women by Race and Ethnicity: Chicago and Urban United States	171
Table 6.2	Employment and Unemployment Distribution of Chicago's Men and Women by Race, Ethnicity, and Poverty Rate	175
Table 6.3	Perceptions of Job Opportunities by Current Neighborhood Poverty: Chicago Men and Women by Race and Ethnicity	179

Table 6.4	Cumulative Proportion of Time at Work Since Age 18 and Mean Number of Jobs Held: Chicago and Urban U.S. Women by Race and Ethnicity	185
Table 6.5	Cumulative Proportion of Time at Work Since Age 18 and Mean Number of Jobs Held: Chicago and Urban U.S. Men by Race and Ethnicity	186
Table 6.6	Average Wages and Means of Support: Chicago Men and Women by Race, Ethnicity, and Current Employment Status	208
Table 7.1	Summary of Racial and Ethnic Effects (Relative to Whites) on Key Outcomes	220
Table 7.2	Percent of Respondents Who Agree That Each Factor Is Very Important for Getting Ahead in Chicago: Men and Women Residing in Chicago's Inner City by Race and Ethnicity	227
Table A1	Distribution of Chicago Census Tracts, 1980 and 1990, by Poverty Rates	240
Table A2	Social and Demographic Characteristics of Chicago Census Tracts, 1980–90	242
Table B1	List of Independent Variables Included in the Analyses	246
Table B4.1	Definitions of Specific Variables Analyzed in Chapter 4	255
Table B4.2	Logit Estimates of High School Noncompletion by Age 19: Men and Women in Chicago and Urban United States	256
Table B4.3	Means of Variables Used to Predict High School Dropout Probabilities: Women and Men in Chicago and Urban United States	257
Table B4.4	Log Odds of Pathways to Family Life by Age 19: Chicago and Urban U.S. Women	258
Table B4.5	Log Odds of Pathways to Family Life by Age 19: Chicago and Urban U.S. Men	259
Table B4.6	Means of Variables Used to Predict Pathways to Family Life Probabilities: Women and Men in Chicago and Urban United States	260
Table B5.1	Definition of Specific Variables Analyzed in Chapter 5	261

Table B5.2 Logit Estimates of Recent Welfare Use: Women Residing in Chicago's Inner City and Urban United States 262
Table B5.3 Logit Estimates of Recent Welfare Use: Men Residing in Chicago's Inner City and Urban United States 263
Table B5.4 Means of Variables Used to Predict Recent Welfare Use Probabilities: Women and Men in Chicago and Urban United States 264
Table B6.1 Definitions of Specific Variables Analyzed in Chapter 6 266
Table B6.2 Logit Estimates of Labor Force Participation: Men in Chicago and Urban United States 268
Table B6.3 Logit Estimates of Labor Force Participation: Women in Chicago and Urban United States 269
Table B6.4 Means of Variables Used to Predict Labor Force Participation Probabilities: Women and Men in Chicago and Urban United States 270
Table B6.5 Probit Estimates of Labor Force Participation: Men and Women Residing in Chicago's Inner City 271
Table B6.6 (Log) Wage Regressions for Men and Women Residing in Chicago's Inner City 272
Table B6.7 Comparison of Actual, Expected, and Reservation Wages: Men and Women Ages 18–44 in Chicago's Inner City by Ethnicity in 1987 273

FIGURES

Figure 2.1 Residential Concentration of Black Population, 1970 36
Figure 2.2 Residential Concentration of Black Population, 1990 37
Figure 2.3 Residential Concentration of Hispanic Population, 1970 38
Figure 2.4 Residential Concentration of Hispanic Population, 1990 39
Figure 2.5 Changes in UI-Covered Manufacturing Employment by Place: Chicago Metro Area, 1970–90 50

Figure 2.6	Changes in UI-Covered Nonmanufacturing Employment by Place: Chicago Metro Area, 1970–90	51
Figure 2.7	Spatial Distribution of Neighborhood Types, 1970	60
Figure 2.8	Spatial Distribution of Neighborhood Types, 1980	61
Figure 2.9	Spatial Distribution of Neighborhood Types, 1990	62
Figure 3.1	Questions Analyzed from the Social Opportunity Survey	75
Figure 4.1	Risk of High School Dropout by Race, Ethnicity, and Sex: Chicago and Urban United States	104
Figure 4.2	Risk of High School Dropout by Family Structure and Sex: Chicago and Urban United States	106
Figure 4.3	Risk of High School Dropout by Family Poverty Status and Sex: Chicago and Urban United States	107
Figure 4.4	Marriage Pathway to Family Life: Chicago and U.S. Urban Women by Race and Ethnicity	110
Figure 4.5	Marriage Pathway to Family Life: Chicago and U.S. Urban Men by Race and Ethnicity	111
Figure 4.6	Birth Pathways to Family Life: Chicago and U.S. Urban Women by Race and Ethnicity	112
Figure 4.7	Birth Pathways to Family Life: Chicago and U.S. Urban Men by Race and Ethnicity	113
Figure 4.8	Marriage Pathway to Family Life: Chicago and Urban Poor U.S. Women	114
Figure 4.9	Racial and Ethnic Differences in Pathways to Family Life by Age 19: Chicago and Urban U.S. Women and Men	116
Figure 4.10	Family Structure Differences in Pathways to Family Life by Age 19: Chicago and Urban U.S. Women and Men	118
Figure 4.11	Family Poverty Status Differences in Pathways to Family Life by Age 19: Chicago and Urban U.S. Women and Men	119

Figure 4.12	Adjusted Household Income by Family Experience: Chicago and Urban U.S. Women and Men	121
Figure 4.13	Welfare Participation by Family Experience: Chicago and Urban U.S. Women and Men	124
Figure 4.14	Adjusted Household Income by Family Experience and Race: Chicago and Urban U.S. Women and Men	125
Figure 4.15	Welfare Participation by Family Experience and Race: Chicago and Urban U.S. Women and Men	126
Figure 5.1	Risks of Recent Welfare Use by Race and Ethnicity: Chicago and Urban U.S. Women and Men	140
Figure 5.2	Risks of Recent Welfare Use by Family Structure: Chicago and Urban U.S. Men and Women	142
Figure 5.3	Risks of Recent Welfare Use by Family Poverty Status: Chicago and Urban U.S. Men and Women	143
Figure 5.4	Risks of Recent Welfare Use by High School Graduation Status: Chicago and Urban U.S. Men and Women	144
Figure 5.5	Risks of Recent Welfare Use by Pathways to Family Life: Chicago and Urban U.S. Men and Women	145
Figure 5.6	Risks of Recent Welfare Use by Current Marital Status: Chicago and Urban U.S. Men and Women	146
Figure 5.7	Risks of Recent Welfare Use by Employment Status: Chicago and Urban U.S. Men and Women	147
Figure 5.8	Duration of Welfare Spells by Type of Aid and Sex	152
Figure 5.9	Duration of AFDC Welfare Spells: Chicago Women by Race and Ethnicity	154
Figure 5.10	Duration of AFDC Welfare Spells: Chicago Women by Pathways to Family Life	155

Figure 6.1 High School Completion Effects on Labor Force Participation: Chicago and Urban U.S. Men and Women 190
Figure 6.2 Experience Effects on Labor Force Participation: Chicago and Urban U.S. Men and Women 192
Figure 6.3 Racial and Ethnic Differences in Labor Force Participation: Chicago and Urban U.S. Men and Women 193
Figure 6.4 Work Exit Rates: Chicago and Urban U.S. Men 197
Figure 6.5 Work Exit Rates: Chicago and Urban U.S. Women 199
Figure 6.6 Waiting Time to Reemployment by Reason for Job Exit: Jobless Chicago Men and Women 201
Figure 6.7 Waiting Time for Reemployment: Chicago and Urban U.S. Men and Women 202
Figure 6.8 Waiting Time to Reemployment: Chicago Men and Women by Race and Ethnicity 205
Figure 6.9 Actual and Predicted Wage Ratios of Minority Groups Relative to Whites, by Employment Status and Sex 210
Figure 7.1 Percent of Respondents Who Agree That Each Factor Is Very Important to Getting Ahead in Chicago 231

Preface

We deliberately chose our title, *The Color of Opportunity,* because *opportunity* is more powerful than neutral synonyms for color and opportunity—*diversity* and *chance.* The term *diversity* cannot convey the continuing weight of ascription in contemporary American society, and *chance* suggests a random process that is somehow beyond the scope of policy instruments. But public policy does influence opportunity—sometimes in planned but more often in unintended ways. Our goal is not to conduct a policy analysis of family, welfare, or employment policy, but rather to inform the policy debates with solid empirical research. In that light, recent poverty and welfare legislation appears to be substantially misguided. The results of our study also do not square with the idea that a robust economy lifts all boats or that two-year time limits on welfare assistance are justified for all groups even when aggregate labor markets are tight. We show that both race and place continue to structure opportunity at the turn of the twenty-first century.

The Urban Poverty and Family Structure study was initially conceived as an ethnography of poor Chicago neighborhoods, but evolved into a multimethod, multiyear study of parents residing in poor neighborhoods. The initial study plan was prepared in 1980 as a component of a larger proposal submitted by NORC in response to a national competition for the Institute for Research on Poverty, housed at the University of Wisconsin since the War on Poverty. A 1983 revision, described as a "modest, hands-on project inspired by the work of William Foote Whyte," was submitted to the Department of Health and Human Services in response to a request for proposals on inner-city neighborhoods using qualitative methods. For political reasons, this study was not funded. The 1984 revision, which was designed as a three-year project that called for a survey of black couples and single men and women as well as an ethnography of black inner-city communities, became the centerpiece of the Chicago Urban Poverty and Family Structure study (see Krogh 1991).

Two circumstances galvanized support for the Chicago project in 1984. First, Charles Murray's (1984) publication of *Losing Ground* rekindled scholarly debate about the causes of persistent poverty, especially about whether public assistance encouraged out-of-wedlock childbearing and undermined labor market activity. Second, Wilson and Neckerman's (1986) male marriageable pool hypothesis, which claimed that the rising share of black single parents paralleled the ratio of employed black males to black females, provided a plausible alternative to Murray's argument about how welfare undermines incentives to work and marry. These provocative and innovative ideas were brought to the attention of the scholarly community when Wilson and Neckerman presented their influential paper in 1984 at a national policy conference sponsored by the Institute for Research on Poverty, entitled "Poverty and Policy: Retrospect and Prospects."[1]

Having mobilized academic and political interest in his project, coupled with strong encouragement from the foundation community, Wilson expanded the scope of the Urban Poverty and Family Structure Project to include whites, Mexicans, and Puerto Ricans residing in poor neighborhoods. The Ford Foundation was particularly instrumental in urging the inclusion of both Mexicans and Puerto Ricans in the study. At a time when the term *underclass* was becoming synonymous with urban blacks, the comparative design promised more convincing statements about how minority group membership shapes opportunity for inner-city residents. The Urban Poverty and Family Structure Project conducted the first survey of inner-city residents, thereby permitting systematic comparisons among blacks, whites, Mexicans, and Puerto Ricans.[2]

Although the Urban Poverty and Family Life Survey (UPFLS) was deliberately designed to understand the family formation, employment, and

1. Coincidentally, Charles Murray published *Losing Ground* the same year. The conference volume was published by Harvard University Press, 1986, as *Fighting Poverty: What Works and What Doesn't,* S. Danziger and D. Weinberg, eds.

2. The HUD Moving to Opportunity surveys and the Multi-City Surveys spearheaded by the Russell Sage Foundation now permit such comparisons, but the thrust of these surveys differs, and not all cities include the four groups.

welfare behavior of poor inner-city residents, ironically, the study has been criticized for adopting this focus. That is, some critics allege that it is not possible to learn much about poverty and welfare behavior by studying poor people without a comparison of the nonpoor.[3] According to critics, a sample of poor people will impose serious limits on generalizability and render statistical analyses of various correlates of poverty (including welfare utilization and pervasive joblessness) problematic. We explain in chapter 3 that the UPFLS did *not* sample poor people, but poor places, and we elaborate the implications of this distinction for the analyses we conduct and the inferences we draw. Moreover, as we demonstrate in chapter 2, Chicago's inner-city neighborhoods, particularly those where poor whites and poor Hispanics are numerically dominant, are socially and economically heterogeneous. Nevertheless, we are careful not to generalize about all urban poor based on a survey of poor inner-city Chicago neighborhoods. Analyses based on the National Survey of Families and Households permit inferences about the urban poor in general, and they serve as a benchmark for understanding the behavior and circumstances of the inner-city poor.

3. These criticisms were articulated at two national conferences. The first was held at Northwestern University (19–21 October), which focused on *The Truly Disadvantaged,* Wilson's seminal book on concentrated urban poverty. Two years later the Chicago Urban Poverty and Family Life Conference was held at the University of Chicago to mark the formal conclusion of the study (10–13 October 1991). Technically speaking, critics maintained that the survey was fraught with serious selection bias because respondents were sampled on the basis of their poverty status. But this attribution is incorrect because the sample was based on poor places, not poor people.

Acknowledgments

This project represents several years of collaboration at the University of Chicago, Stanford's Center for Advanced Study in the Behavioral Sciences, Tel Aviv University, the Rockefeller Foundation's Research and Conference Center at Bellagio, the Catholic University of Nijmegen in the Netherlands, and Princeton University. The publication of this book brings closure to work begun *in* Chicago, where we met in 1987 as student and professor, and our work *about* Chicago. This enterprise is quintessentially Chicago, inspired by our esteemed colleague, W. J. Wilson, who generously gave us access to the Urban Poverty and Family Life Survey data even before he had finished his own writing and provided unequivocal encouragement throughout our journey. Publication of this monograph marks the completion of a decade-long collaboration, which not only produced several empirical papers, but also and more important, consolidated one of the richest friendships to which two academics could ever aspire. Our ability to complete this project amidst changing locations bears witness to our resolve to overcome the challenges of distance, time, and family obligations.

This project began with a small grant of ten thousand dollars in 1989 from the Rockefeller Foundation for an analysis of welfare participation based on the Urban Poverty and Family Life Survey. The opportunity to spend a year at the Center for Advanced Study in the Behavioral Sciences (1990–91) allowed us to write several empirical papers about the employment experiences of Chicago's inner-city fathers and to plan a book-length treatise about Chicago's urban poor. Although we began planning the monograph a decade ago, the story line was vague, and the color and opportunity theme was ill defined. In fact, the "story" about the color of opportunity did not emerge until we began decoding racial and ethnic differences in pathways to family life and their attendant consequences for economic well-being. This phase of our work was made possible by Guggenheim and Sackler fellowships, which permitted Tienda to spend three months in Tel Aviv in early 1994. We are grateful for the leave time these fellowships provided, including the supplement from the University of Chicago's Division of Social Sciences. There we crafted the "Spouses and Babies" paper (*Ethnic and Racial Studies,* 1997), which was pivotal not only in clarifying the story line but also in providing clues about how a life-course perspective connected the three empirical chapters under the color and opportunity theme.

Several "writing camps" in Chicago (summer 1996), in Bellagio, Italy (summer 1997), and in Princeton, New Jersey (January 1999 and November 1999), were necessary to bring our work to fruition. Summer camp in Chicago was a crucial turning point in the production of the monograph. We completed a first draft of the empirical analyses for all chapters, and drafted chapter 2 with the assistance of Jeff Morenoff. Summer camp at Bellagio was the second key milestone in our voyage, because there we strengthened the theoretical underpinnings of the monograph, sharpened the research questions, and set a plan to revise and finalize the empirical analyses. We gratefully acknowledge funding from the Rockefeller Foundation for this splendid opportunity amid inspiring environs. The revisions were completed over the following year with the help of a sabbatical from Tel Aviv University and a fellowship to Haya from the Catholic University of Nijmegen. Tienda's relocation to Princeton slowed progress, as did a broken hand the day before our semifinal "writing camp" in January 1999, but we were too close to shore to abandon ship. The manuscript was completed in July 1999 with excellent technical assistance from Pamela Bye-Erts and Adair Iacono, who patiently revised multiple drafts of tables, figures, and chapters.

Although we did not receive a dedicated grant to conduct this research, we have benefited from the institutional support provided by our universities, especially the Population Research Center at the University of Chicago, the Office of Population Research at Princeton University, and the Social Sciences Division of Tel Aviv University. In addition, support from the Russell Sage Foundation as part of a larger study about the transition from school to work permitted us to cover what little direct costs were not borne by our institutions.

Finally, we wish to acknowledge many colleagues who contributed immeasurably to our mission. Our primary thanks are to Bill Wilson, former colleague and special friend, who believed in our enterprise even before we had a clearly charted course. Richard Taub provided several hands-on lessons about Chicago by bringing to life the neighborhoods that were merely census tracts in the statistical analysis. His profound knowledge of the "City of Big Shoulders" added substance to our understanding of the empirical results we report in chapter 2. Jeff Morenoff prepared the maps, assisted with the statistical analysis of Chicago neighborhood transitions, and prepared the descriptive tabulations about social and demographic change from 1970 to 1990. We are deeply indebted to Russell Hardin, who read the entire manuscript in a very rudimentary form; his probing comments and suggestions significantly improved the final manuscript. In addition, many of our colleagues at Chicago, Tel Aviv, and Princeton helped sharpen the final product as we presented ideas at conferences and seminars. At the Chicago Urban Poverty and Family Life Conference held in October 1991, Ron Mincy commented on the first draft of our analyses about color and employment opportunity that appear in chapter 5, and Mary Jo Bane challenged us to be explicit in acknowledging the strengths and limitations of the UPFLS design, especially in studying welfare participation. All names used for verbatim quotations are fictitious, but many names represent special people in our lives.

Our deepest debt, of course, is to the men in our lives—Rami, Asaf, and Omer (H. S.); Luis and Carlos (M. T.). Their patience and tolerance of prolonged separations for our "writing camps," and their willingness to undertake disruptive transcontinental trips to support our endeavor enabled us to bring this project to fruition. Now when Carlos asks, "Did you finish it again?" we can answer yes, and this is the final time!

Chapter One

Poor People, Poor Places

As we searched for an appropriate introduction to our title, *The Color of Opportunity,* it occurred to us that the most powerful voice is not our own but those of inner-city parents who inspired and directed our ideas. They may not be certified social scientists, but they understand all too well the relation between color and opportunity. Take Gina, a white female we interviewed. When queried whether the United States was a land of opportunity, she replied most candidly:

> *Sometimes it seems like . . . there are a lot of opportunities, but only certain people are allowed to get 'em. Sometimes like with different jobs and stuff, you have to be . . . you have to know somebody to get a . . . promotion, or you have to be the right color or the right uh . . . last name or whatever, and I don't think that makes it such a land of opportunity for people that aren't those things.*

Letoya, a black female respondent, sees it differently: "There are better opportunities for white people. Say, upper-class and middle-class white people. But lower-class white people, they have a hard time as black people do." But Manuel, an unemployed father of six with only nine years of education, is far more optimistic, perhaps because he compares his situation with alternatives in Mexico. In his words,

> *I agree that it's the land of opportunity. To the bone. Yes sir. There's no other country like this one. . . . Oh yes, yes. Sure. When you say this is the land of opportunity, you get more than anywhere else with less effort. It's getting kind of hard now, but it's still better than other countries.*

However, many unskilled workers do not share his optimism, particularly blacks like Sabrina, who feel crowded out of the labor market. She says,

> *They are bringing too many foreigners over here from other countries for any of us to get ahead or do what it's our right to do, and that's put us down. . . . I don't believe it's just in the neighborhood, I believe it's true for the whole United States. This is not the only lower-class neighborhood; we have these all over. (emphasis added)*

These diverse views about economic opportunity are from parents who resided in Chicago's inner city during the late 1980s. Their responses to questions about opportunity instantiate long-standing academic and policy debates about the causes and consequences of urban poverty. During the 1980s and early 1990s social scientists identified several features that allegedly changed the character of poverty: its spatial concentration; its chronicity among a subset of the poor; its involvement of more women and a broader ethnic spectrum; and the heightened salience of structural factors in perpetuating inequality (W. J. Wilson 1987; Jencks and Peterson 1991).

The strong economic performance and tight labor markets of the mid- to late 1990s deflected academic discourse about concentrated and persistent urban poverty and redirected the interest of the social science research community to the paradox of rising inequality during periods of sustained economic prosperity (Danziger and Gottschalk 1995). Yet persisting poverty remains an important social policy challenge because the economic fortunes of the disadvantaged have declined in absolute and real terms since the mid-1970s, and poverty remains unacceptably high for the most affluent nation in the world. Although employment growth has been impressive during the latter 1990s, less skilled and low-paid workers—those in the lowest quartile of the earnings distribution—remain in serious economic trouble.[1]

Given these mixed economic messages and allegations that the inner-city poor, who live in concentrated poverty neighborhoods, differ from the urban poor in general, we address empirically several questions left unanswered by the urban underclass debate that dominated academic and policy discourse from the mid-1980s until the mid-1990s. First, do the inner-city poor differ from the urban poor in their family, welfare, and

1. In 1996 the median income for all families was $42,300, but the medians for whites, Hispanics, and blacks were $44,756; $26,179; and $26,522, respectively. Half these medians, which is the standard for measuring relative poverty, represents monthly incomes of slightly over $1,000 (U.S. Department of Commerce 1998, table 746).

employment behavior, and are there racial and ethnic differences in these behaviors?

In the main we find few differences between the inner-city poor and all urban dwellers, and even fewer between the inner-city poor and the urban poor in general. Although rates of joblessness and welfare receipt are higher among Chicago's inner-city residents compared to all central-city residents, and even to the urban poor, these differences largely derive from the more limited opportunities for earning a living in Chicago following the city's massive job declines after 1973. We find substantial trace-group differences in labor force and welfare participation rates; however, once factors that influence these outcomes are modeled statistically, racial and ethnic differences largely disappear. This is because blacks, whites, Mexicans, and Puerto Ricans differ systematically in various characteristics that render them less competitive as workers. In particular, early experiences with poverty and low human capital investments carry over into adulthood and undermine workers' earning capacities, and their ability to qualify for income transfers.

Precisely because groups differ in their early experiences with material disadvantage and acquisition of human capital, our second goal is to examine the circumstances that give rise to unequal educational attainment and work experience from a life-course perspective. In this way we address how decisive early disadvantages are in determining the well-being of minority and nonminority parents. Both nationally and in Chicago, we find large racial and ethnic differences in family structure and childhood exposure to poverty and welfare receipt, with blacks considerably more likely than either whites or Hispanics to have been reared by a lone parent or to have experienced material deprivation during childhood. Also, Chicago parents are more likely than all urban parents, including poor urban parents, to have been poor while growing up. This is socially and economically significant because it increases their risks of becoming poor adults. We identify two mechanisms that perpetuate economic and social disadvantage over the life course, namely, premature withdrawal from school and nonmarital childbearing.

Like many prior researchers, we find persisting racial differences in family formation patterns, with black women significantly more likely than either whites or Hispanics to remain unmarried and to bear children out of wedlock. However, this pattern obtains among all urban women, especially the urban poor, and cannot be identified as a ghetto-specific behavior that distinguishes the inner-city poor from the poor in general. Accordingly, we conclude that both race and poverty independently influence the likelihood of entering family life via a birth relative to the more normative route of marriage. That this pathway to family life has lasting consequences for economic well-being is socially significant because it undergirds racial and ethnic differences in adult poverty.

The final issue we consider is why welfare rates are so high and, correspondingly, employment rates so low in Chicago's inner city. More specifically, we evaluate whether racial and ethnic differences in welfare use and labor force participation are due to lack of marketable skills, to unequal labor market opportunity, or to preferences for welfare rather than work. Analyses of welfare participation revealed appreciably higher rates of current and recent use in Chicago compared to all urban areas, and small differences between Chicago parents and the urban poor nationally. In general, racial and ethnic differences were similar across settings. The notable exception is the higher and more chronic welfare use among blacks in Chicago, which we argue reflects the sorting processes that segregate blacks into the poorest neighborhoods rather than ghetto-specific behavior or race-specific preferences for welfare. In fact, we show that the vast majority of welfare recipients prefer a job to public assistance and, quite surprisingly, that white, jobless, inner-city parents are less willing than jobless minorities to accept a job. We argue, therefore, that higher welfare participation rates and lower employment rates in Chicago reflect the more limited job opportunities there compared to central cities nationally.

The life-course perspective provides important insights about the factors responsible for racial and ethnic variation in work and welfare behavior because it illustrates how early disadvantages accumulate in the transition to adulthood. We show that the large differences in labor force activity of black, white, Mexican, and Puerto Rican men are produced by unequal stocks of human capital and work experience. The acquisition of work experience is a key factor in racial and ethnic differences in labor force status among mature adults. Once groups are statistically standardized according to characteristics that influence work behavior, racial and ethnic differences in labor force participation disappear in Chicago, although some remain among urban poor residents nationally. Thus, to the extent that minority-group status influences economic opportunity, it does so largely through pivotal life-course events that undermine life chances during adulthood. However, we do find some evidence of wage discrimination: minority workers who do manage to find employment receive lower wages than their similarly skilled white counterparts.

The purpose of this chapter is to frame in theoretical terms how minority status is related to family structure, welfare dependence, and labor force behavior. Because it rekindled discussions about the salience of culture and structure in perpetuating poverty and economic disadvantage, the urban underclass debate summarized next provides a useful point of departure. The final section recapitulates the discussion and orients the subsequent analyses by identifying how the themes of color and opportunity are explored in the empirical analyses.

The Urban Poverty Debates

The resurgence of interest in poverty during the 1960s, and again during the 1980s, shared concerns with family disintegration, nonmarital childbearing, welfare dependence, and detachment from the world of work. Another dominant theme in both periods was the disproportionate representation of blacks among the urban poor. Finally, both periods debated the relative importance of cultural and structural forces in driving the growth of urban poverty. Yet, several features of the academic and policy debate during the 1980s were new: references to the urban underclass; avoidance of references to poverty as a cultural syndrome; the idea that concentrated urban poverty differs from poverty in general both in its causes and its consequences; and, with the rapid growth of the poor Hispanic population in the largest urban centers, questions about whether the urban underclass was exclusively black (Moore 1989).

Hindsight shows that the poverty controversies of the 1960s and 1980s are more similar than different, especially regarding the unproductive and ideological debate over the relative importance of the structural versus cultural underpinnings of poverty (Valentine 1968; W. J. Wilson 1991, 1987). Both generations of poverty researchers shared concerns about the chronicity and intergenerational transmission of non-normative behaviors associated with material deprivation, notably welfare dependence and growth of mother-only families. Stimulated by the intense criticism surrounding publication of *The Declining Significance of Race* (1979), W. J. Wilson (1987) generated several new ideas about how structural factors perpetuate concentrated urban poverty. He claimed (1) that selective out-migration of the black middle class was responsible for the growth of a residentially concentrated black underclass; (2) that being poor in a ghetto poverty neighborhood was qualitatively different from being poor in a working-class neighborhood; and (3) that the rise of mother-only families among blacks was directly related to changes in the employability of black men because jobless men were unattractive marriage partners.

W. J. Wilson's (1987) influential treatise on the urban underclass was framed as a response to his critics about the declining significance of race. To that end, he focused on black poverty and introduced the idea of "concentration effects" to theorize in structural terms why mother-only families, chronic joblessness, pervasive welfare use, and other non-normative behaviors were so pervasive in high-poverty neighborhoods. Attempting to explain the dramatic rise in mother-only families among blacks, Wilson proposed the male marriageable pool hypothesis; to explain the rise of concentrated urban joblessness, Wilson focused on the decimation of job opportunities in central cities, coupled with the selective out-migration of relatively more employable families—usually whites. However, Wilson did not consider whether these experiences were shared by other minority groups. Significantly, the Hispanic presence increased dramatically in many cities that experienced massive industrial transformation and job loss precisely when the economic status of blacks was deteriorating.

The growing Hispanic presence in Chicago brought to light several anomalies that might qualify Wilson's assertions about the relative importance of declining job opportunities for the rise of concentrated black poverty. Why did unskilled Mexican immigration to Chicago increase during the 1970s and 1980s, precisely when declining employment opportunities drove both whites and blacks from the inner city? If declining job opportunities were the main cause of rising black joblessness, why were Mexicans with less education able to find jobs in the same labor market? More generally, how applicable are Wilson's explanations to Hispanics and other immigrant minorities? Are his ideas specific to urban blacks?

Although Hispanic poverty also rose during the 1980s, it differed from black poverty in several important ways. First, Hispanic poverty was less concentrated than black poverty; second, increases in child poverty rather than adult poverty largely drove the rise in Hispanic poverty overall; third, trends in Hispanic poverty differed among national origin groups both in cause and intensity because working poverty is more pervasive among Mexicans, whereas nonworking poverty prevailed among Puerto Ricans; and finally, Hispanic poverty was less responsive to economic recovery during the 1980s and early 1990s compared to black poverty (Moore 1989; Tienda 1995; Tienda and Jensen 1988). Recent data indicate that Hispanic poverty actually increased by three percentage points between 1990 and 1996, while black poverty declined by a comparable amount

(U.S. Department of Commerce 1998, table 756).[2] These differences between blacks and Latinos, coupled with the fact that Hispanic national origin groups are residentially concentrated in different regional labor markets, render the task of evaluating the influence of color (i.e., group membership) and opportunity (i.e., employment, poverty, and marriageability) more difficult. However, they also increase the value of Chicago as a case study because it is one of the few labor markets where Mexicans, blacks, and Puerto Ricans compete with whites for economic opportunities. Chicago is an important case study for another reason, namely, its role in inspiring Wilson's early and later ideas.

In his more recent book, *When Work Disappears* (1996), Wilson reiterates several key contentions from his earlier work, namely, that irregular work activity and labor market detachment, which are often associated with chronic welfare participation, distinguish the inner-city poor from the poor in general. Mindful that economic inequality increased during the 1980s and early 1990s, despite strong economic performance, Wilson explicitly acknowledged the significance of race in determining access to economic opportunity in his more recent writings. Nevertheless, these revisions do not alter Wilson's earlier premises that chronic joblessness, labor force detachment, and pervasive welfare participation distinguish the urban underclass from the poor in general, and that declining job opportunities are primarily responsible for the rise of concentrated urban poverty. Conservatives interpreted rising and concentrated poverty to mean that the working poor differ behaviorally from the nonworking poor and that chronically jobless adults are *unwilling* to work, preferring welfare or illegitimate sources of income to legitimate work. With relatively few exceptions, these issues have not been scrutinized empirically, partly because census-type data limits the types of questions that can be addressed, and partly because few studies systematically compared residents of urban *poor places* with representative samples of the *urban poor.*

Wilson's hypotheses about the structural factors undergirding the rise of concentrated poverty neighborhoods prompted a social science industry to test these hypotheses.[3] Two general types of investigations drove the

2. Black and Hispanic poverty rates converged in 1994 for the first time, and since then Hispanics have remained worse off on the indicator of economic well-being, vigorous growth notwithstanding.

3. The new generation of poverty researchers enjoyed several advantages over their predecessors, notably the availability of longitudinal data well suited to evaluating issues of chronicity and intergenerational transmission mechanisms. Also, advances in multilevel modeling increased the possibility of establishing contextual effects on behavioral outcomes.

new poverty agenda: those focused on identifying the origins and perpetuation of urban ghettos, and those focused on the mechanisms responsible for rising poverty among blacks (Jencks and Peterson 1991). Unfortunately, the distinction between poor people and poor places was conflated in many discussions of minority urban poverty, often, ironically, in attempts to separate the structural from the cultural underpinnings of concentrated minority poverty (Tienda 1991). A sample of poor people would not allow us to draw inferences about whether and how the poor differ from the nonpoor in their attitudes and behavioral dispositions. However, sampling poor places, especially those where the majority of residents are not poor, not only allows such inferences, but also permits us to compare poor people across neighborhoods that differ in their concentration of poverty.

Most discussions of the mechanisms fueling the rise of concentrated urban poverty presume that economic disadvantages are accumulated over time. Massey and Denton (1993) made this point explicit in their sharp criticism of urban poverty researchers for failing to recognize the profound influence of residential segregation both in restricting opportunities for blacks and in spatially concentrating poor blacks as poverty rates increased. Correspondingly, the distinction between "reshuffled" and "event-caused" poverty forced researchers to reconsider the mechanisms driving changes in economic well-being over time and across generations. Specifically, Mary Jo Bane and David Ellwood (1986) pointed out that in most cases poverty associated with dropping out of school, becoming unemployed, or bearing a child as a teen was not *caused* by these events, since the individuals involved were already poor when these events occurred. Bane and Ellwood's argument implies that the negative outcomes are not the cause of poverty, but rather one mechanism through which poverty is transmitted intergenerationally.

Bane and Ellwood's (1986, 22) assessment of large racial differences in event-caused versus reshuffled poverty challenged researchers to specify further what factors produce alternative pathways to poverty, and what kinds of experiences are conducive to more or less frequent episodes of poverty. The life-course approach brings into sharp focus the salience of early experiences with material deprivation in shaping trajectories of adult disadvantage among residents of poor, inner-city neighborhoods. Documenting racial and ethnic differences in the perpetuation of economic disadvantage is a *central objective of our study.* Accordingly, we elaborate this theme theoretically by examining the links between group membership, family structure, and welfare use, as well as how the accumulation of disadvantage produces unequal labor market experiences along racial and ethnic lines.

The *life-course perspective,* which focuses on the timing and sequencing of key events and transitions, provides a general framework for conceptualizing how poverty is transmitted across generations (Elder 1994; Hogan and Astone 1986). Developmental psychologists working from this perspective emphasize cognitive functioning and behavioral dispositions that unfold in social and biological time. W. J. Wilson's (1987) formulation of the "social buffer" hypothesis is entirely consistent with this perspective. Wilson argued that exposure to working adults fosters attitudes and behavior toward work that are crucial for understanding whether youth become sufficiently attached to the labor market to embark on productive adult careers. As we show in subsequent chapters, early exposure to economic deprivation reproduces disadvantages through various life-course transitions that have lasting consequences in adulthood and direct implications for economic well-being, especially when slack labor markets limit job options for the unskilled.

Within the life-course perspective, two theoretical constructs are particularly germane for understanding why some young men and women develop the capacity for economic independence, and others do not. One is Sampson and Laub's (1993) notion of a *turning point,* which they define as "a complex conjunction of events that 'turns' a career toward or away from mature normative behavior." Examples include bearing a child out of wedlock, graduating from high school, or, for welfare mothers, marrying. For instance, the occurrence of a nonmarital birth before age twenty can limit employment and income opportunities well into adulthood. Similarly, failure to complete high school may preclude adequate labor market attachment during young adulthood, particularly in an economic climate that rewards the most highly educated workers to the detriment of those with less education.

Pathway is a second construct derived from the life-course perspective that provides theoretical means for analyzing the events conducive to stable work careers. Following Hogan and Astone (1986), we use the term *pathway* to connote a sequencing of life-course events that has implications for future transitions and outcomes. We build on these constructs to address how early investments (education) and pathways to family life lead to self-sufficiency and stable labor market careers during adulthood. In the following sections we elaborate on these themes to identify clues about how color circumscribes opportunity.

Family Structure and Poverty in Life-Course Perspective

Although marriage followed by childbirth remains the normative order of life-course events, changes in nuptiality and reproductive behavior during the last fifty years have diluted the traditional norm. Equally important, large racial and ethnic differences in propensities to marry, to divorce, and to bear children out of wedlock have rekindled discussions about whether and how cultural factors shape marriage preferences and tolerance for nonmarital fertility. Instead, Wilson's male marriageable pool hypothesis emphasized the lack of economic opportunity as the culprit for high rates of nonmarital childbearing among blacks (W. J. Wilson and Neckerman 1986; W. J. Wilson 1987). For Wilson, poverty rather than cultural preferences undergirds racial and ethnic differences in family formation patterns.

The *timing* and *sequencing* of marriage relative to births, which define alternative *pathways* to family life, have enduring consequences for mothers' and children's well-being (Stier and Tienda 1997). This is because the timing and sequencing of both events affects socioeconomic status, completed fertility, marital stability, psychological well-being, and intergenerational continuities in family structure and poverty (McLanahan and Sandefur 1994; McLanahan and Booth 1989; Upchurch and McCarthy 1990; D. Anderson 1993). Moreover, deleterious consequences associated with early and out-of-wedlock fertility often persist, even if women marry subsequent to having a child, or divorce after having been married. For example, never-married mothers are more likely to be welfare-reliant and unable to maintain steady work compared to married women who postpone childbearing until they are able to support a child.

In other words, poverty both perpetuates and is perpetuated by nonmarital fertility because women who enter family life by having a baby are more likely to be from poor families, and their children are more likely to be poor (McLanahan and Sandefur 1994; Bane and Ellwood 1986). Compared to children reared in two-parent families, those reared in parent-absent families are more likely to withdraw from school prematurely, to become teen parents themselves, to develop weak labor market attachment, and to become economically dependent on public income transfers. These intergenerational relationships persist even after controlling for human capital and other characteristics that influence employment prospects and welfare eligibility.

McLanahan and Sandefur (1994) identified several mechanisms portraying how family structure affects the risk of poverty and economic well-being. First, families headed by single parents, usually mothers, experience more economic deprivation and financial instability. Low economic resources are linked with a host of problem behaviors among adolescents. Transgressive behavior is related to the limited community resources to engage adolescents in productive activities. Second, because the family is the primary context for early socialization through role modeling, the behavior of the custodial parent conveys values toward family life. For example, owing partly to reduced parental supervision, being reared by a single mother can foster higher tolerance for out-of-wedlock childbearing. Third, time invested in childrearing is usually compromised by the absence of a parent, and a reduction in parental supervision is associated with transgressive behavior and poor scholastic performance (Hogan and Kitagawa 1985; McLanahan and Sandefur 1994). Reduced parental investments in children are especially harmful if poor single parents must also assume domestic financial obligations.

References to race-specific family formation patterns during the 1980s were highly reminiscent of the "culture of poverty" thesis prominent during the 1960s (see reviews by Taylor et al. 1990; and Vega 1990).[4] However, the more enlightened formulations recognized explicitly that family structure is both a vehicle for the transmission of poverty and a consequence of being poor (Bane and Ellwood 1986; McLanahan and Sandefur 1994). Several studies show that women reared in mother-only families were more likely to become single mothers themselves and to receive public assistance (Hogan and Astone 1986; McLanahan and Booth 1989; McLanahan and Sandefur 1994).

That the strength of these linkages differs among blacks, whites, and Hispanics raises questions about whether economic deprivation alone or group-specific behavioral dispositions toward marriage and childbearing are responsible for racial and ethnic variation in family structure. Because poor women are more likely to become single mothers and black women are more likely to be poor, the higher rates of single motherhood among blacks may be due to their poverty, not their race or race-specific preferences for spouses or babies. Alternatively, high rates of unmarried teenage childbearing may derive from "ghetto-specific" behavior; that is, they may

4. Baca Zinn (1989) summarized the cultural and structural interpretations of the link between family structure and poverty, but her characterization of the studies conducted in the context of the underclass debate was too simplistic because she ignored numerous studies that acknowledged the importance of *both* cultural and structural forces. Moreover, by uncritically accepting the structural model and its narrow implications, she also failed to theorize how cultural factors may shape differences in minority experiences with poverty.

be a direct consequence of concentrated urban poverty (Hannerz 1969; Testa et al. 1989; E. Anderson 1991; Massey and Denton 1993; W. J. Wilson 1996). This view parallels the culture of poverty thesis that non-normative family formation patterns represent behavioral adaptations to economic deprivation (Lewis 1966, 1961; Stack 1974; Rodman 1971).[5] Unfortunately, the preoccupation of the underclass debate with the experiences of urban blacks deflected attention from the rise in mother-only families among Hispanics, especially among Puerto Ricans.

Nevertheless, because minority women are more likely than whites to have been reared by single mothers and to have been reared in poverty, it is difficult to disentangle the influence of family structure, poverty, and minority status on family formation patterns. Stated differently, research about the family formation patterns of the inner-city poor is unclear as to whether negative outcomes associated with premarital fertility largely reflect their occurrence *to unmarried women;* to *minority mothers; to poor mothers;* or the *combined* impact of these factors. Yet this is precisely the distinction needed to support claims about the emergence of ghetto-specific and race-specific behavior (see Patterson 1993; W. J. Wilson 1987; Hannerz 1969; E. Anderson 1991).

Chapter 4 argues that *both* minority group status and early life experiences with poverty exert independent effects on the likelihood of family formation through births versus marriage. The existing empirical literature provides little guidance about whether the experiences of black and Hispanic women are similar in this regard. Wilson's thesis about the urban underclass implies that residents of ghetto poverty neighborhoods differ from minority populations and the poor in general with respect to their marriage and fertility behavior. However, this issue has not yet been subjected to systematic empirical scrutiny. Because most poverty researchers either analyze nationally representative data (e.g., the Panel Survey on Income Dynamics) or conduct ethnographic research on a single neighborhood or small group of families (E. Anderson 1991; Hannerz 1969; Stack 1974), it is impossible to determine whether and in what ways the behavior of minority populations residing in and outside of high-poverty neighborhoods differs.

In chapter 4 we address this limitation of prior research by analyzing

5. In fact, these discussions are quite similar to the "culture of poverty" argument that was popular during the 1960s. Although the original formulation of this thesis was based on ethnographies in Puerto Rico and Mexico (Lewis, 1966, 1961, 1959), direct references to these populations were conspicuously absent in discussions of poverty and family structure during the 1980s.

a representative sample of parents residing in poor neighborhoods and a nationally representative survey of urban parents. Our empirical tasks include documenting racial and ethnic differences in *pathways* to adult family life; ascertaining the relative importance of early deprivation (opportunity) versus minority group status (color) on *pathways* to family life; and evaluating the consequences of entering family life via birth or marriage. Specifically, we consider whether the pathways to family life are significantly different among inner-city mothers versus urban mothers generally, as implied by the growing references to "ghetto-specific" behavior.

Welfare Participation in Life-Course Perspective

In the wake of the Great Society programs, social science research on welfare participation burgeoned. Analysts documented trends in the use of various programs, evaluated work incentives, and attempted to answer seemingly simple questions about the behavioral determinants of welfare participation and the duration of spells. Most research focused on Aid to Families with Dependent Children (AFDC) because it was the largest means-tested income transfer program; because it is associated with vague conceptions of the undeserving poor (unlike Supplemental Security Income, which is designated for the "deserving" poor, i.e., the aged, blind, and disabled poor); and because politicians became increasingly worried about the rise in the AFDC caseload after 1965 (Moffit 1992; Blank and Ruggles 1993). Social scientists investigated whether this growth was due to changes in eligibility criteria, disincentives to work associated with expanded benefit packages, or changes in behavior among a growing "welfare class" defined by chronic dependence over generations (Rein and Rainwater 1978; Duncan, Hill, and Hoffman 1988).[6] In the public and academic debate, chronicity and widespread dependence in poverty neighborhoods were the core issues.

Although the total AFDC caseload increased over time, in fact, AFDC *participation rates* declined after 1973 both because of changes in eligibility criteria and because of changes in the propensity to accept welfare among benefit-eligible women (Moffit 1992; Blank and Ruggles 1993). Because participation rates are contingent on eligibility, increases in the caseloads were fueled by the growth of the benefit-eligible population, namely, mother-only families. This revelation refocused attention on the

6. Before Auletta (1982) popularized the term *underclass*, Rein and Rainwater (1978) used the term *welfare class* to focus on the fact that welfare recipients do not work.

circumstances driving changes in family structure, notably the growth of households headed by single women. Both poverty and the welfare system itself were scrutinized as primary transmission mechanisms.

On the heels of a deep recession and in a political environment hostile to poor people, Charles Murray's book, *Losing Ground* (1984), was highly influential not only because it reinforced prevailing stereotypes about poor minority women as members of a welfare class, but also because it held the welfare system accountable for perpetuating economic dependence. Murray argued that the growth of AFDC caseloads could be traced to two perverse incentives inherent in the program design: (1) less restrictive eligibility criteria (namely, the elimination of restrictions on mothers rights to cohabit with men to whom they were not married); and (2) increased generosity of benefits. Addition of food stamps and Medicaid to the support packages of poor, single women with dependent children rivaled what they could earn in the labor market (Edin and Lein 1997). Thus Murray reasoned that changes in the generosity of AFDC benefits undermined incentives to marry and work while encouraging nonmarital childbearing, fostering prolonged dependence, and perpetuating poverty over generations. However baseless empirically, these claims resonated with the political agenda of conservatives anxious to reform, if not abolish, AFDC, as they did in 1996.[7]

The debate surrounding Murray's position rekindled academic interest in questions about intergenerational transmission mechanisms during the early 1980s. In the 1960s, Lewis asserted that the culture of poverty was transmitted intergenerationally via the internalization of deviant values and norms. Although much academic debate ensued (see Valentine 1968), few attempted to evaluate these claims empirically. However, during the 1980s, claims that the welfare system discouraged marriage and encouraged nonmarital fertility were subjected to extensive empirical scrutiny using longitudinal data. Most studies focused on questions of incentives (i.e., What aspects of the benefit package encourage dependence and discourage work?), but several highly influential studies directly examined

7. With the passage of the Personal Responsibility and Work Opportunity Reconciliation Act of 1996 (PRWORA), public assistance in the United States changed fundamentally. The new law shut down several federal programs, including Aid to Families with Dependent Children (AFDC), and replaced them with block-grant welfare funding known as Temporary Assistance for Needy Families (TANF). While the target program recipients—single mothers and their children—remain the same, the new program creates incentives to reduce the welfare roles, obliges states to move recipients into work activities, and imposes time limits on welfare receipt (Blank 1997). Our study focuses on AFDC because the survey predates TANF by a decade.

questions of chronicity and intergenerational transmission mechanisms (Bane and Ellwood 1986; O'Neill, Bassi, and Wolf 1987; Plotnick 1983; Duncan, Hill, and Hoffman 1988).

On questions of AFDC participation, turnover, and the dynamics of welfare utilization, consensus emerged that (1) utilization rates were higher among older, less educated, less healthy women and those from larger families; (2) black women were more likely to participate than whites of similar educational and age composition; and (3) more generous benefits extended the duration of spells (Duncan and Hoffman 1988; Moffit 1992). Virtually all studies that examined participation and turnover found that the benefit level was negatively and significantly related to the probability of leaving AFDC and positively related to the probability of entry.

Equally important for our research interest in AFDC receipt are consistent findings that demographic events, notably changes in family composition, were more influential than labor market events in determining welfare participation (Bane and Ellwood 1986; O'Neill, Bassi, and Wolf 1987; McLanahan 1988, 1985; Tienda 1990). The consensus among researchers was that about three out of four new welfare entries were triggered by changes in family composition—45 percent due to changes in marital status, and 30 percent due to teenage fertility (Duncan and Hoffman 1988). Welfare exits also were driven by demographic events, namely, marriage (35 percent) and changes in eligibility associated with children's departures from the family (11 percent).

Two additional findings are germane to our interest in racial and ethnic differences in welfare transmission mechanisms over the life course. One is that the durations of welfare spells differ depending on *pathways* to entry, with a crucial difference between women who enter an AFDC episode via out-of-wedlock birth and those who enter because of marital disruption. A second key finding of prior studies is the persisting racial difference (net of education and age) in welfare behavior, both in participation rates (Blank and Ruggles 1993; Moffit 1992) and in the duration of welfare spells (Duncan and Hoffman 1988).[8] In contrast to the 1960s, when racial differences were equated with "cultural"

8. Mark Rank (1994, 1988) did not find racial differences in the duration of welfare spells, but his analysis did not take into account differences in propensity to participate, nor did he consider paths out of welfare. Also, his crude measurement of duration renders his analysis crude relative to investigations based on nationally representative surveys with greater precision in the measurement of welfare episodes.

differences, persisting racial effects have not been explained in such terms except via linkages to large differences in marriage behavior (Bane and Ellwood 1986) and the growth of mother-only families (Moffit 1992; McLanahan and Garfinkle 1989; McLanahan and Sandefur 1994).

Despite growing consensus about the importance of demographic events in accounting for the dynamics of welfare participation, apprecia-ble controversy remains about the determinants of AFDC receipt, partly because the growth of mother-only families was itself attributed to the welfare system. According to this logic, demographic events that trigger spells of AFDC use (i.e., divorce and nonmarital fertility) were themselves influenced by the generosity of welfare benefits and hence contributed to the growth of the welfare-eligible population. In fact, studies that consider whether the welfare system was responsible for the growth of mother-only families and increases in nonmarital fertility found weak and inconsistent evidence (e.g., cf. Bahr 1979 and Draper 1981; see also Cain 1985). Moreover, studies that investigate whether *tastes* for public assistance are cultivated and transmitted intergenerationally were inconclusive. The weight of evidence linking AFDC to family structure showed "no measurable impact on births to unmarried women and only a modest effect on marriage, divorce, separation, or the proportion of female-headed families" (Duncan and Hoffman 1988). However, questions about the behavioral effects of the welfare system, in particular, whether prolonged use had a destructive effect on personal values and attitudes, remained unsettled.

Supporters of more generous benefits emphasized the repeated finding that the majority of new AFDC spells last no more than two years (Rein and Rainwater 1978; Bane and Ellwood 1986, 1983; Plotnick 1983). Yet ample evidence existed that a subset of the welfare population (about 10 percent) experienced spells that lasted over ten years. For those episodes, the probability of leaving AFDC rolls decreased the longer the spell lasted. This evidence reinforced stereotypes about the emergence of a welfare class and fueled political concerns about the chronicity of dependence and its intergenerational transmission.[9]

9. Moffit (1992) observed that declining exit rates with increasing length of welfare spells (negative duration dependence) was consistent with the idea that self-esteem declines as episodes of welfare receipt increase (see also Popkin 1990); that the stigma associated with welfare declines as women become accustomed to their poverty lifestyles (see also Horan and Austin 1974); that a protracted episode serves as a negative signal to employers (Kirschenman and Neckerman 1991); and that propensity to remain on welfare for a long period may result from characteristics unobserved (i.e., unmeasured) by survey researchers, including preference for welfare over work, but also from constraints of having to survive on benefits that declined in real terms (see, for example, Edin and Lein 1997).

Whether daughters of welfare mothers are more likely than daughters of nonwelfare users to become welfare recipients remains one of the most challenging questions about the determinants of welfare behavior. This is because a positive association between mothers' and daughters' welfare use could be produced through direct and indirect effects, each of which implies very different transmission mechanisms and different policy implications. *Direct effects* imply that being socialized into welfare generates a culture of dependence that is transmitted intergenerationally. Evidence that growing up in a welfare family altered young girls' preferences for welfare—by lowering the stigma associated with receipt, by transmitting information about eligibility criteria, or by lowering the transaction costs associated with securing a grant—would constitute a *direct* effect on the duration of a welfare spell. *Indirect* effects arise because poor mothers use welfare benefits, which in turn influences daughters' propensity to receive welfare because of choices that reduce earnings capacity during adulthood, notably educational underinvestment and premarital childbearing (McLanahan and Sandefur 1994; Moffit 1992; Gottschalk 1990). Unfortunately, separating the direct from the indirect effects has proven an intractable analytical problem.[10]

Empirical studies that used panel data to investigate this question show strong and consistent positive associations between parental welfare use and daughters' participation in means-tested transfer programs (Moffit 1992). But it is not obvious that such linkages represent a change in preferences for welfare, because the positive intergenerational correlation also reflects the transmission of economic disadvantages associated with being reared in a female-headed family, in a low-income family, or both (McLanahan and Sandefur 1994). In fact, evidence that welfare participation is *directly* related to the growth of an urban underclass is very limited (Duncan and Hoffman 1988), although several analysts have conceded that the concentration of poverty is conducive to non-normative behavior that includes chronic welfare participation (W. J. Wilson 1996, 1991; Massey and Denton 1993).

According to Rein and Rainwater (1978), the proper comparison to determine whether welfare has the effect of perpetuating itself in the second generation is between low-income parents who went on welfare and

10. Because it is causally prior to the welfare participation decision, modeling the intergenerational transmission of poverty could give some purchase on distinguishing the direct from the indirect effects. However, this problem imposes serious endogeneity problems because similar processes govern parental and offspring's behavior.

low-income parents who did not. Their comparisons revealed that children of welfare recipients were no more likely than children of nonrecipients to participate in means-tested income programs themselves. From a life-course perspective, this implies that welfare participation decisions reflect the intergenerational transmission of poverty both through the economic disadvantages reproduced from having been reared by a poor single parent, and through the compromising choices about school completion, family formation, and labor market activity that ensue (McLanahan and Sandefur 1994; An, Haveman, and Wolfe 1993; Duncan and Hoffman 1988). In other words, not all children reared in poverty become welfare-dependent single mothers who fail to complete high school, but the likelihood of such outcomes is certainly greater for those who are reared in poverty, especially in the absence of role models pursuing more normative family formation and employment behavior.

In sum, a life-course perspective on welfare participation can be fruitfully engaged to explore questions about transmission mechanisms, specifically, how racial and ethnic differences in welfare participation reflect disadvantaged beginnings, alternative pathways to family life, and the associated consequences of early life-course decisions, including educational investments and labor market attachment. Given the apparently strong linkages between the economic circumstances of the family of orientation and family of procreation, combined with strong empirical evidence that demographic events are pivotal turning points shaping trajectories of welfare participation, we examine in chapter 5 whether men and women of color who experience similar opportunities to redirect their socioeconomic lives are equally likely to become welfare-reliant.

Labor Market Behavior in Life-Course Perspective

In the context of the underclass debate, attempts to explain pervasive joblessness in inner-city neighborhoods inferred that irregular work activity and chronic joblessness distinguished the inner-city poor from the poor in general. Presumably, when skilled and semiskilled jobs diminished as a share of total employment in old industrial centers (Kasarda 1995, 1985; Holzer 1991), residents of impoverished neighborhoods became detached from the labor market because fewer role models and social networks were present to establish connections with the world of work (W. J. Wilson 1996, 1987); because illicit income opportunities offered more lucrative alternatives to low-wage jobs (Freeman 1992); and because the generosity of welfare benefits undermined the value of conventional employment (Murray 1984; Freeman 1992). Whether defined as a behavioral syndrome associated with extreme income deficiencies (Auletta 1982) or as a set of spatial and social arrangements that situated individuals outside the social mainstream (W. J. Wilson 1987), weak labor force attachment became another important marker that allegedly distinguished the urban underclass from the poor in general. By extension, members of the urban underclass were presumed unwilling to work, preferring welfare or illegitimate sources of income over legitimate work. With relatively few exceptions (e.g., Edin and Lein 1997; Tienda and Stier 1996b, 1991), this question was not systematically addressed empirically.

Labor market attachment is an important marker of future economic success, and it is tightly interwoven with previous life-course events, notably educational attainment and family status. Future earnings and economic well-being can be traced to the timing and quantity of educational investments, the accumulation of labor market experience, and, particularly for women, the timing of first births and marital dissolution (Ahituv, Tienda, and Tsay 1998; Tienda and Stier 1996a; Ben-Porath 1967; Willis 1986). To understand racial and ethnic differences in labor market outcomes over the life course, we examine the timing of investments in human capital, the pervasiveness and nature of discrimination, and how changing family statuses constrain labor market opportunities. We elaborate each of these in turn.

On the timing of human capital investments, a vast empirical literature has demonstrated not only the value of acquiring labor market skills *early* in the life course, but also how decisions about the timing of school departure, family formation, and labor force entry can have lasting consequences for economic well-being (Willis 1986; Mincer 1974; Becker 1964; Ben-Porath 1967; McLanahan and Sandefur 1994). However, recent trends in educational attainment and employment of minority adult men present an interesting puzzle that a life-course perspective of human capital investments might clarify. That is, Mexicans and Puerto Ricans average fewer years of school than either whites or blacks, yet they participate in the labor force at higher rates than blacks and average higher wage rates (U.S. Department of Commerce 1993; Greenwood and Tienda 1998). Furthermore, it appears that racial and ethnic labor market inequality begins quite early in the life course, because Hispanic youth drop out of high school at much higher rates than either black or white youth, but black youth confront greater obstacles in securing jobs whether or not they graduate from high school (Hotz and Tienda 2000; Ahituv, Tienda, and Hotz 1999). This paradox suggests that early labor market experiences may be pivotal for understanding the processes that generate profiles of disadvantaged adults, and it highlights the promise of a life-course perspective on labor market experiences for understanding the emergence of racial and ethnic differences in adult labor market status.

Racial and ethnic differences in human capital accumulation also derive from life-cycle experiences with joblessness and unstable employment. The significance of employment instability for understanding high rates of inner-city joblessness stems from a simple empirical finding, namely, that early employment experiences, while not perfectly deterministic, strongly influence future labor market outcomes.[11] This seemingly obvious point is methodologically difficult to establish—particularly when the goal is to link a *sequence* of experiences that occur over the life course to an outcome at a particular point in time. The classic status-attainment framework in sociology, for example, cannot reveal whether particular *sequences of events* (i.e., prolonged or frequent joblessness, intermittent school attendance, or the occurrence of a birth or marital disruption) are pivotal in producing profiles of labor market disadvantage among young and mature adults.[12] Frequent transitions to joblessness, which are more common among less-educated and less-experienced workers, not only increase the likelihood of permanent labor force withdrawal, but also reduce the accumulation of valuable work experience over the life course (Hsueh and Tienda 1995, 1994; Tienda and Stier 1996b). Acquisition of work experience is particularly important for individuals with low stocks of formal schooling because it represents a form of human capital that makes workers valuable to employers (Mincer 1974; Devine and Keifer 1993). Given assumptions that inner-city workers are less committed to the labor market, it is surprising that discussions of inner-city poverty have largely ignored the role of unstable employment in producing profiles of chronic joblessness and economic disadvantage.

The theoretical distinction between general and specific training also informs the life-course perspective of work careers (Becker 1993, 33–51).

11. See Clogg, Eliason, and Wahl (1990) for an eloquent statement of this relationship and supporting evidence based on the Current Population Surveys. Hsueh and Tienda (1996) have extended these ideas using the Survey of Income and Program Participation.

12. First-generation analyses of the socioeconomic life cycle made powerful assumptions in modeling intragenerational processes using a single indicator of early labor market experience (usually first full-time job) with current outcomes. See Duncan, Featherman, and Duncan (1972), and Featherman and Hauser (1978). Current empirical applications employ dynamic choice models to relate life-cycle experiences to subsequent outcomes. See Hotz et al. (1999) and Ahituv, Tienda, and Tsay (1998).

General training, which is usually equated with formal and vocational schooling, involves the acquisition of skills that are transferable across firms. Excepting employer-subsidized educational programs or government-sponsored training programs, individual workers usually bear the costs of acquiring general training. Also, to maximize returns over the life course, investments in general training typically are made at younger ages (Ben-Porath 1967). Narrowly defined, specific training refers to skills whose productivity benefits are exclusive to firms that provide the training. Because much training acquired in the workplace is neither completely specific nor completely general, years of work experience captures all forms of human capital accumulation that enhance workers' attractiveness to employers beyond the appeal of their general skills (education) while also providing signals about workers' commitment to jobs.

That minority and female workers experience more labor force instability than men and nonminority workers is highly relevant to our interest in how racial and ethnic inequality is produced over the socioeconomic life cycle (Clark and Summers 1979; Akerlof and Main 1981; Clogg, Eliason, and Wahl 1990; Hsueh and Tienda 1994; F. D. Wilson and Wu 1993). Concretely, this implies that group differences in early educational investments *and* in accumulated labor force experience are at least partly responsible for racial, ethnic, and sex differences in labor market outcomes. Among women, racial and ethnic differences in labor force participation may also be produced by group differences in the occurrence and timing of births, which greatly constrain labor market options among poor women (Ahituv, Tienda, and Tsay 1998).

Group differences in accumulation of general and specific human capital over the life cycle might clarify why racial and ethnic differences in labor market outcomes widen at successive ages (Ahituv, Tienda, and Hotz 1999), but they do not necessarily explain why comparably educated black, inner-city residents are less likely than whites or Hispanics to be offered a job in the first place (Tienda and Stier 1996b; 1991). Mexicans' greater success in securing jobs and commanding higher wages than comparably educated blacks may reflect their early acquisition of valuable labor market experience (which employers may take as a signal of market commitment), or it may result from employers' discriminatory preferences for Mexican over black workers. However, average racial and ethnic differences in labor force participation rates and wages may also result from group differences in *returns* to human capital—what is often interpreted as discrimination. Few studies have provided direct evidence of discrimination, but workers' perceptions of disparate treatment coupled with evidence of unequal employment and wage outcomes among comparably skilled workers would be more convincing than either source of evidence alone. In chapter 6 we argue that *discrimination* remains a powerful force producing racial and ethnic wage inequality, that it affects workers over their entire life course, and that it is buttressed by more restricted opportunities for earning a living among younger cohorts relative to more mature workers.

Discrimination can be understood both in individualistic and structural terms. Conventional "structural" explanations for the rise of inner-city joblessness focus almost exclusively on the disappearance of well-paid manufacturing jobs in old industrial centers (see W. J. Wilson 1987; Kasarda 1995, 1985). Wilson's proposal that the decline of manufacturing jobs from inner cities was responsible for chronic labor market inactivity in ghetto-poverty neighborhoods does not resolve the puzzle of large racial and ethnic differences in employment status within a *single* labor market.[13] Although seldom explicit, most discussions of labor market discrimination infer both perspectives, which are embodied in the notion of statistical discrimination—the idea that disparate treatment of any given individual is based on perceptions or stereotypes of an entire group.

Queuing theory provides another structural variant of how discrimination produces racial and ethnic labor market inequality because group outcomes are tied to changing opportunities, whether cyclically or structurally produced (Hodge 1973). From the perspective of queuing theory, when job opportunities requiring low to moderate skills (education) decline, as they have since the mid-1970s, ethnic employment competition intensifies because employers can be more "choosy" about which workers to hire. Moreover, when labor markets are slack, employers' preferences and prejudices may play themselves out differently than they would in a situation of labor scarcity (Hodge 1973). Furthermore, and this is relevant for understanding the emergence of racial and ethnic differences within a single labor market, demographic changes that change the ethnic composition of prospective workers may also alter the nature of job competition along color lines. For Chicago, Tienda and Stier (1996a, 1996b, 1991) showed that Mexican immigrants have higher labor force participation

13. Although his argument about social isolation is provocative, Massey and Denton's (1993) forceful analysis of how residential segregation produces concentrated urban poverty is more convincing about the role of race in delimiting social and economic opportunity. As we document in chapter 2, in Chicago, poor blacks are far more residentially segregated than either Mexicans or Puerto Ricans, and this outcome was produced by racial discrimination in housing markets, which in turn delimited educational opportunities and labor market options.

rates and receive lower wages than comparably skilled blacks.[14] This finding does not deny the importance of declining opportunities for earning a living, but it underscores that both color (group membership) and opportunity (availability of jobs commensurate with workers' skills) are relevant for understanding the evolution of racial and ethnic stratification generally, and profiles of extreme disadvantage in particular.

Finally, our concerns about the emergence of structured inequalities over the life course include the origin of sex differences in labor force activities. It is well known that the female labor supply is more sensitive than male market activity to family events such as birth, marriage, and divorce. Also, racial and ethnic differences in female labor supply are well documented, even if less well understood. However, from a life-course perspective, the *timing* of family events (especially the occurrence of a birth during adolescence or outside of marriage) has profound implications for women's labor market prospects—much more so than similar events do for men.

Our premise that racial and ethnic labor market inequality begins early in the life course does not necessarily require identical mechanisms for young men and women. For men, most recent work has focused on the school-to-work transition, but there is less comparable information for young women (see review in Donahoe and Tienda 2000). For women, most research on early life-course transitions has emphasized how adolescent childbearing compromises investments in education, while neglecting the repercussions of early labor force experiences for long-term economic independence (Ahituv, Tienda, and Tsay 1998). Given the importance of labor market experience as a form of human capital and as a signal about labor market attachment, racial and ethnic differences in women's work experience are filtered through several mechanisms: (1) those that originate because of alternative pathways to family life (which are more consequential for women than for men); (2) those that accumulate over the life course (because women with many children are particularly susceptible to spells of labor market withdrawal for family reasons and thus often acquire less work experience); and (3) those that emerge because women's work is valued less than men's in the labor market.

14. Several studies have documented that immigrants participate in the labor force at a higher rate than natives (see Smith and Edmonston 1997 for the most comprehensive review to date). Aggregate econometric evidence indicates that the labor market impacts of recent immigrants are relatively benign, but studies of specific firms and communities reveal intense competition between immigrants and native minorities (see Rosenfeld and Tienda, 2000).

In summary, a life-course perspective provides a strong foundation for understanding both racial and ethnic differences in labor market outcomes of parents residing in poor places whose opportunities to find work differ because of minority group membership; because of geographical differences in the availability of jobs; because of disadvantages transmitted from the family of orientation to the family of procreation; because of early family life experiences, especially the premature assumption of childcare responsibilities, which undermine the accumulation of human capital necessary for ensuring productive work careers; and because gender continues to define very different socioeconomic options for poor men and women. Chapter 6 systematically compares the labor force careers of both men and women residing in poor urban neighborhoods with their racial and ethnic urban counterparts nationally.

Synthesis and Overview

We have argued that the life-course perspective provides a useful framework for investigating the emergence, the contours, and the consequences of racial and ethnic inequality among parents who live in poor neighborhoods. In chapter 4 we describe racial and ethnic differences in *pathways* to family life and analyze their consequences for future earnings capacity and economic self-sufficiency. However, in keeping with our notion of turning points, we investigate whether the deleterious effects of entering family life via birth rather than marriage can be moderated by subsequent behavior. For example, are welfare participation rates lower and family income levels higher among teenage mothers who finish school or eventually marry? Or do the deleterious consequences of alternative pathways to family life persist even among those who attempt to turn around their lives by entering into traditional family arrangements and investing early in marketable skills? Do parents residing in poor places differ from all urban poor parents in their pathways to family life and subsequent economic well-being? We document clear racial and ethnic differences in pathways to family life that are *not* unique to parents residing in poor places but emerge among urban parents nationally. In chapters 5 and 6, respectively, we investigate how disadvantages accumulated over the life course—from family of orientation to family of procreation—influence the welfare participation and employment experiences of minority parents. We consider how the early life-course experiences of black, white, Mexican, and Puerto Rican parents compare with respect to critical events (turning points) that have decisive and lasting repercussions on their adult functioning. We also evaluate whether the linkages between early life experiences and subsequent adult outcomes are uniform among demographic groups in order to understand why some groups (e.g., blacks) appear less able to transcend poverty and its consequences.

That blacks are more likely than Mexicans or Puerto Ricans to live in the poorest neighborhoods renders more difficult the task of separating the effects of race from spatially circumscribed differences in opportunity, which often manifest as distinct social norms (Massey and Denton 1993; Tienda and Stier 1991). Selective comparisons of parents who reside in Chicago's poor inner-city neighborhoods with nationally representative samples of urban parents and with poor urban parents matched on race and Hispanic origin provide some leverage to overcome this methodological challenge, but our evidence on "neighborhood effects" is indirect at best. Like other social scientists, we are unable to document the sorting and selection processes that produced Chicago's high-poverty neighborhoods in the first place except in broad historical terms (Tienda 1991; Massey and Denton 1993; Jargowsky and Bane 1991). This is the subject of chapter 2, which frames our study of color and opportunity in Chicago in historical, demographic, and economic terms.

Our approach differs in several important respects from most studies of the urban underclass that consider whether and how the inner-city poor differ from the poor in general. Although the core analyses we present are based largely on a single case study, systematic comparisons with national urban dwellers broaden the lessons about racial and ethnic inequality beyond Chicago's city limits. Chicago is an ideal case study to investigate how economic opportunity is delimited for minority populations. As the most racially segregated city in the country, it provides an extreme case of spatially circumscribed social and economic resources and thus sets outer limits for any claims about race and place in producing inequality among minority and nonminority groups. That is, because of the extreme residential and economic segregation of the city, any effects of race and place are likely to be more pronounced in Chicago compared to other large metropolitan areas.

Furthermore, the social transformation of Chicago's inner city as a result of industrial restructuring was the basis of Wilson's (1987) ideas that rekindled the poverty debates during the 1980s. Thus, spotlighting Chicago allows us to evaluate empirically Wilson's theoretical ideas in terms of the setting that inspired them. The Urban Poverty and Family Life Survey, which we analyze, was designed explicitly to address several questions about the incidence and persistence of behaviors that allegedly distinguish the inner-city poor from the poor in general. In particular, the study sought to characterize the lives of residents from poor neighborhoods, but the sampling frame fortuitously included neighborhoods with poverty rates of 20 percent or higher. While this includes the poorest ghetto-poverty neighborhoods—the focus of the underclass controversy—it also includes neighborhoods with substantial numbers of working poor. This design permits us to compare the family, work, and welfare behavior of parents residing in neighborhoods that differ in their density of poor people. Thus, a second unique feature of our study stems precisely from the rich data on which our empirical analyses are based. These data, although based on a cross-sectional survey, contain several retrospective histories that permit us to reconstruct the life course of respondents residing in neighborhoods that represent diverse social and economic environments.

Throughout we emphasize the distinction between *poor people* and *poor places,* which was confounded in many studies of concentrated urban poverty during the 1980s. To be sure, sampling in poor neighborhoods increases the likelihood of selecting poor people, but in a highly income- and race-stratified city like Chicago, many nonpoor residents live in poor areas, and most poor do not live in ghetto-poverty neighborhoods. For the Urban Poverty and Family Life Survey, which sampled neighborhoods with poverty levels above 20 percent, this fact simply means that many nonpoor respondents were included in the sample. By systematically comparing Chicago's inner-city residents to a national sample of central-city residents and to a nationally representative subset of *poor* central-city residents, we are able to assess the relative importance of race and place in governing the behavior of the urban poor.

Our study expands the concept of color to include Mexicans and Puerto Ricans along with blacks and whites, and place variation includes diversity in neighborhood poverty levels as well as cross-city variation in the national urban landscape. This design is essential to address numerous questions that were not settled by prior studies, which either studied only the poor or assumed that underclass membership was predominantly a black phenomenon. In the absence of comparative evidence, the presumption that blacks residing in ghetto-poverty neighborhoods behave differently from the poor in general (because they experienced prolonged joblessness and welfare dependence and were detached from the world of work) became reified as fact. Our analyses go beyond existing studies in clarifying the intersection of race and class by showing that the experiences of Chicago's inner-city parents are less atypical when compared to the urban poor nationally.

Finally, our study of urban poverty is unique in its life-course approach and its consideration of several outcomes that bear on and reflect economic well-being. Our approach to understanding family formation, welfare participation, and employment experiences emphasizes group differences in the accumulation of disadvantages throughout the life cycle, and the empirical analyses focus on pivotal events—turning points—that have important consequences for subsequent outcomes. Although it is convenient for analysts of poverty to concentrate on contemporary outcomes, such as current welfare participation or employment status, we demonstrate throughout the importance of situating current statuses relative to early experiences. Many poor parents were poor children, and these circumstances limited their opportunities to acquire the tools and life skills they needed to compete in a rapidly changing labor market. Thus, the superficial lack of racial effects in welfare participation and employment behavior can be largely traced to our explicit representation of the accumulation of disadvantages over individuals' life courses. That is, given that minorities, but especially blacks, are more likely than whites to have been reared in poverty by a lone parent, then the adult consequences associated with these background factors, mediated by pivotal life events, explain the disproportionate concentration of minorities among jobless welfare participants.

Chapter Two

Chicago: Economic and Social Transformation of an Urban Metropolis

Hog Butcher for the World,
Tool maker, Stacker of Wheat
Player with Railroads and the Nation's Freight Handler;
Stormy, husky, brawling;
City of Big Shoulders

—Carl Sandburg, "Chicago"

Inspired by his working-class background and penchant for social criticism, Carl Sandburg, in *Chicago Poems,* portrays the harsh realities of urban life generally and in the industrial Midwest specifically. Chicago offered ample opportunity for those willing to labor in the slaughter houses, steel mills, railroads, and manufacturing plants, but the price was long and arduous working conditions, ethnic prejudice, and extreme social inequality. Like New York, Pittsburgh, and Detroit, Chicago epitomized urban-industrial society at the turn of the twentieth century, which symbolized both economic progress and the loss of community that urban sociologists would soon come to lament (Wirth 1938; Fisher 1995). Although Sandburg's admiration for the working-class spirit is evident in our epigraph, his "Chicago poems" deplore the failures of the American Dream. The themes of ethnic prejudice, class inequality, and the drudgery of work undergird Sandburg's personification of a city vibrant with opportunity, yet burdened with social injustice.

Were he a contemporary observer, Sandburg would find striking continuities between Chicago at the beginning of the twentieth and the twenty-first centuries. Group membership qualified social and economic opportunity in Chicago in both periods. Race and ethnic competition over jobs and social positions remains a defining feature of Chicago's social history. According to Drake and Cayton (1993, 17),

> the entire history of Chicago, from its birth to the First World War, was characterized by the struggle, sometimes violent, of the first-comers and

This chapter was written in collaboration with Jeffrey D. Morenoff.

> native-whites against the later immigrants—the "foreigners." In the [1860s] it was everybody against the Irish and the Irish against a handful of Negroes and Hungarians; in the [1870s] it was the Know-Nothing native-Americans against the Germans, Irish, Scandinavians, Bohemians, Slavs, and Frenchmen; in the [1890s], the northern Europeans and native-whites against the southern and eastern Europeans, the so-called "new immigration."

This history is very much the same at the turn of the twenty-first century, but the actors have changed. Rather than southern and eastern Europeans, Latinos, Asians, African Americans, and a dwindling white ethnic immigrant population occupy center stage in Chicago's ethnic landscape.

Both continuity and change mark the history of social and economic inequality in Chicago. While color and class continue to shape access to social and economic opportunity, Chicago has also witnessed profound demographic and socioeconomic transformations that have reshaped its racial and ethnic landscape, its class structure, the ecology of jobs, and the very character of its neighborhoods. One example of such change is Chicago's declining population base coupled with shifts in its racial and ethnic composition, which has rendered whites a clear minority of the city population.[1] Another is the demise of Chicago's traditional industrial employment base, which has meant fewer and generally less well-paid jobs in the city as compared to the surrounding metropolitan area since 1970. Changes in population and employment are the pillars on which we develop our arguments about how color, that is, group membership, delimits opportunity in Chicago.

Accordingly, this chapter frames our study of family life, welfare participation, and employment in Chicago's inner city by describing how economic opportunity declined while racial and ethnic inequality rose. The chapter is divided into three parts. The first section provides an abbreviated demographic history of Chicago, with a focus on recent changes in the city's ethno-racial composition as well as trends and differentials in educational attainment, labor force participation, poverty, and welfare participation. This overview provides a historical context for understanding how economic opportunities withered in the inner city. By tracing the growth and decline of employment for the six-county Standard Metropolitan Statistical Area (SMSA), Cook County, and the City of Chicago, we illustrate the salience and consequences of spatially distributed opportunity. In addition to depicting the multidimensional character of neighborhood growth and decline, this discussion situates the Chicago

1. Whites still outnumber minorities in the surrounding suburban ring.

neighborhoods included in the Urban Poverty and Family Life Survey against the backdrop of social, demographic, and economic changes underway since 1970. The concluding section summarizes main trends and derives implications about the significance of minority-group status for social and economic well-being.

Demographically, we trace Chicago's changing ethno-racial composition to three trends: the exodus of whites from the city to the suburbs; the dramatic growth of the Latino population; and the slow growth of the black population. Ecologically, the social landscape was transformed by the spatial polarization of neighborhoods, which was driven by the dramatic loss of jobs since 1970, coupled with high levels of residential segregation. Our characterization of ecological polarization differs from prior representations by focusing on a multidimensional characterization of inequality rather than simply on poverty and race.

Chicago as Crucible of Diversity and Change

Approximately sixty years after its incorporation in 1830, Chicago's population exceeded one million persons, of which three in four were either foreign-born or of foreign parents (Drake and Cayton 1993, 8).[2] Rapid population growth fueled by immigration from eastern and southern Europe, the pull of industrial jobs, and natural increase doubled the city's population between 1900 and 1940, from 1.7 to 3.4 million. Although World War I slowed the flow of immigrants from Europe, labor shortages precipitated internal migration from surrounding rural areas and triggered the influx of black migrants from the South.

In fact, the black presence in Chicago dates almost to its incorporation as a city, when the Underground Railroad operated to free black men from the oppression of slavery. Internal migration following the abolition of slavery fed the growth of Chicago's black population, which represented a meager 2 percent of city residents at the turn of the twentieth century. However, labor shortages created by the World War I attracted blacks from the South, resulting in a doubling of the black population share to 4 percent by 1920.[3] This was but a dress rehearsal of what would follow

2. The first settlement at what is presently Chicago is dated circa 1790, but the city was not incorporated until approximately 1830 (Drake and Cayton, 1993, 8, 31). The brief historical overview draws from Drake and Cayton's classic, *Black Metropolis,* 1993 edition, which provides an excellent point of departure for interpreting recent social, demographic, and economic trends in the city.

3. Authors' tabulations from *Black Metropolis,* table 1, p. 8.

World War II. Not only did immigration to Chicago slow, but internal migration from the South to the Midwest also accelerated. By the end of World War II, nearly one in ten of Chicago's 3.6 million residents was black.

Two general themes emerge from the numerous volumes that have chronicled the social and economic history of black Chicago. One is that blacks experience the most extreme levels of residential segregation, at least by comparison to other ethnic groups. In the words of Massey and Denton (1993, 8), "Residential segregation is the institutional apparatus that supports other racially discriminatory processes and binds them together into a coherent and uniquely effective system of racial subordination."[4] A second major theme in the history of black Chicago epitomizes the core of our work, namely, how color restricted opportunity. In *Black Metropolis,* Drake and Cayton chronicle in great detail how Chicago's black population was consistently overrepresented among the ranks of the jobless. In addition, among the educated who managed to secure employment, blacks were extremely underrepresented in jobs commensurate with their training. Although urban blacks who migrated north improved their economic condition relative to their southern counterparts, their economic progress lagged far behind that of European immigrants, including the most recent newcomers. Drake and Cayton (1993, 223–24) claim this resulted because advancement was not based on individual merit; because blacks were not allowed to compete on equal footing with white workers; and because rigid job ceilings provided unequal access to moderate- and high-status jobs, effectively truncating black opportunities for social mobility.

Compared to blacks, the Latino presence in Chicago is a twentieth-century phenomenon. Latinos are the most diverse of Chicago's ethnic groups, representing over twenty national origins, but Mexicans have been numerically dominant throughout the twentieth century, and their presence continues to grow (Rosenfeld and Tienda 2000). Mexican

4. Massey and Denton explain that ethnic and immigrant neighborhoods differ in several important respects that clearly distinguish them from the ghettos. Not only are immigrant neighborhoods more heterogeneous, but the total share of any single group that resides in ethnic neighborhoods nowhere approaches the share of the black population that resides in black neighborhoods. Equally important, residence in ethnic neighborhoods and enclaves represented a transitory phase of immigrants' social and economic adaptation experience, but ghettos were enduring and permanent features of black's residential experience. Thus, even when maximally segregated, the social isolation experienced by European ethnic groups was well below that of blacks and was of shorter duration.

immigration to Chicago began around 1910, when the revolutionary upheavals pushed thousands of workers and their families across the Rio Grande. Chicago, then a bustling industrial hub, was an attractive midwestern destination. Recruiters hungry for unskilled labor lured Mexican workers to jobs in the meat-packing industry and the relatively well-paying steel mills, thereby consolidating a visible Mexican presence on the west and south sides of the city (Año Nuevo de Kerr 1975).[5] The influx of Mexican workers to Chicago continued for nearly two decades, and by the time the Great Depression virtually stopped all U.S.-bound international migration, the city's Mexican origin population had reached twenty-five thousand (Casuso and Camacho 1985). Expanding labor market opportunities fueled by World War II and the ensuing economic prosperity rekindled immigration from Mexico. This and high fertility rates fueled the growth of Chicago's Mexican-origin population after 1950, but especially after the 1965 Amendments to the Immigration and Nationality Act were fully operational.

The Puerto Rican presence in Chicago is even more recent, beginning mostly after World War II. Puerto Ricans are the second largest Latino population in the city. Although Puerto Ricans have lived in the continental United States for more than a century, visible Puerto Rican communities emerged during the 1950s when cheap airfare and postwar job growth encouraged mass migration. Although New York City remained the preferred destination, thousands of Puerto Rican migrants headed to the Midwest—primarily Chicago—during the 1950s and 1960s, mostly lured by the promise of stable employment and improved living conditions. Their class background as unskilled laborers put them in direct competition with Mexicans and blacks.

An important difference between the Mexican and Puerto Rican experience in Chicago can be traced to the distinct historical periods in which both migration streams originated and evolved. Chicago's economy was still growing during the 1950s and 1960s, but the economic opportunities that enabled early Mexican immigrants to participate in the American Dream were not available for Puerto Ricans. Mexican immigrants to Chicago found jobs in heavy industry, earned family wages, entered the economic mainstream, and moved to the suburbs with their English-speaking children and grandchildren. That well-paid, unskilled jobs began to de-

5. The main destination areas for Mexican immigrants were communities known as Back of the Yards and South Chicago. For a brief and informative history of the Mexican presence in Chicago, see Año Nuevo de Kerr (1975).

cline just when the Puerto Rican diaspora was unfolding forced the new migrant group into entry-level positions in service and light manufacturing industries (Casuso and Camacho 1985, 2). Thus, compared to Mexicans, the Puerto Rican community made relatively little socioeconomic progress over time, and they became the poorest Latino population in the city, as well as the nation (Bean and Tienda 1987; Tienda and Jensen 1988; Tienda 1989; Casuso and Camacho 1985). That many Puerto Ricans are black or mulatto may also explain why Puerto Ricans fared as poorly as blacks in Chicago. That is, their racial phenotype makes them more vulnerable to discrimination than Mexicans. However, Casuso and Camacho (1985, 13) argue that the poor economic standing of Puerto Ricans also may be traced to the higher economic expectations of the second generation, whose members seem unwilling to take jobs their parents filled. We consider this claim explicitly in chapter 6 and find no support for it.

In summary, the migration of blacks, Mexicans, and Puerto Ricans to Chicago since World War II began a process of ethnic transformation that reconfigured the city's ethnic and socioeconomic landscape. Therefore, in the following sections, we review recent demographic, employment, and poverty trends in order to situate the ensuing plight of inner-city parents against recent sweeping socioeconomic changes in Chicago.

Recent Demographic Trends

World War II not only stimulated job growth, but also fueled the city's population growth to satisfy rising labor demand. Opportunities seemed limitless, and Chicago's population peaked in 1950 at 3.6 million. As the war-driven labor demand began to fall, the city's population began to shrink. Although major population losses from industrial urban centers are generally considered to be a phenomenon of the 1970s and 1980s, in reality gradual population losses had begun a decade or two earlier. As occurred in other northern, industrial cities, Chicago's population began to shrink slowly during the 1950s, coincident with the impetus of suburbanization (Suttles 1990, 24–25). However, Chicago's 2 percent population loss during the 1950s would pale by comparison to the losses sustained during the 1970s and 1980s. Chicago's population fell by more than three hundred thousand (11 percent) during the 1970s, and by an additional two hundred and twenty thousand residents (7 percent) during the 1980s. As Chicago's population fell, the greater metropolitan area, which includes DuPage, Kane, Lake, McHenry, and Will counties in

Table 2.1 Racial and Ethnic Composition of Metro Area: Cook County and Six-County SMSA, 1970–90

	1970	1980	1990
Cook County			
Whites	74.1	43.6	38.2
Blacks	21.4	39.3	38.9
Hispanics			
Mexicans	1.9	8.7	12.4
Puerto Ricans	1.7	3.8	4.2
Others	n/a	1.9	2.4
Asians	0.7	2.3	3.6
Others	0.3	0.5	0.3
N[a]	54,656	75,616	108,266
Six-County SMSA[b]			
Whites	78.7	69.3	73.0
Blacks	17.5	19.9	15.8
Hispanics			
Mexicans	1.6	5.2	5.9
Puerto Ricans	1.4	1.9	1.4
Others	n/a	1.3	1.1
Asians	0.6	2.0	2.6
Others	0.2	0.4	0.2
N	69,451	178,244	392,515

Source: 1970, 1% Public Use Microdata Sample (PUMS); 1980, 5% PUMS; 1990, 5% PUMS
[a] 1990 sample *N*'s unweighted.
[b] Six counties include Cook, DuPage, Kane, Lake, McHenry, and Will counties.

addition to Cook County, grew by 9 percent (Chicago Fact Book Consortium 1992).

The growth of Chicago's suburban communities and adjacent counties, which was driven by selective out-migration of whites, produced major changes in the city's racial and ethnic composition. During the 1970s, about one-third of Chicago's white population left the city. Although the growth of the black population had slowed, the exodus of whites resulted in a rising share of minorities in Cook County and within the city limits. Continued in-migration of Latinos partly offset the loss of white residents and attenuated the rate of population decline, but it also altered the ethnic landscape in decisive ways. As revealed in table 2.1, between 1970 and 1980 whites ceased to be the majority population in Cook County, although they remained dominant in the metropolitan area.

In 1960, Chicago's black population exceeded eight hundred thousand, representing 23 percent of the city population (Hirsch 1983). By 1970, the city's black population exceeded one million; at that time, one in every three Chicagoans living in the inner city was black compared to fewer than 18 percent of the metropolitan-area population (21 percent of Cook County). Two decades later, blacks made up nearly 40 percent of Cook County's population, but only 16 percent of the six-county SMSA. Although the white share of the metro population declined during the 1970s, its share recovered slightly during the 1980s. Concurrently, the black population share of the SMSA fell below its 1970 level.

The Hispanic population residing within the city limits more than doubled between 1970, when approximately 7 percent of all Chicagoans were Hispanic, and 1990, when just under 20 percent were Hispanic (Chicago Fact Book Consortium 1992). The burgeoning Mexican-origin population increased its share of the six-county metropolitan population from 1.6 to 6 percent between 1970 and 1990, with most of this growth occurring during the 1970s. Within Cook County alone, the Mexican population share increased six-fold in twenty years, from 1.9 percent in 1970 to 12.4 percent in 1990. Cook County's Puerto Rican population share rose more gradually, from 1.7 percent to 4.2 percent between 1970 and 1990. But these aggregate changes in population composition conceal the crucial ingredient for understanding how minority-group status influences economic opportunity, namely, residential segregation. Chicago remains the most racially segregated city in the country, despite slight decreases in segregation levels in recent years (Massey and Denton 1993).

The pervasiveness of racial residential segregation remains a defining feature of Chicago's ecological landscape. Figures 2.1 and 2.2 depict the process of ghetto expansion in Chicago from 1970 to 1990. The 1970 map is highly reminiscent of Drake and Cayton's portrayal of the city in 1945, which showed that color lines were well established even before World War II. Running along the South Side of the city, Chicago's "black belt" originally included the communities of Douglas, Oakland, and Grand Boulevard, and by 1960 it extended farther south to communities such as Washington Park, Woodlawn, and Englewood. Chicago's "second ghetto" (Hirsch 1983) involved black settlements along the city's West Side. Figure 2.2 shows that neighborhoods on the periphery of the southern and western corridors of the original black belt experienced the most dramatic change in racial composition. Although the path of ghetto expansion had been laid out well before World War II, the size of the ghetto, its spatial contiguity, and its social marginality changed appreciably. As we demonstrate below, contemporary segregation is far more egregious in its social and economic consequences than it was before.

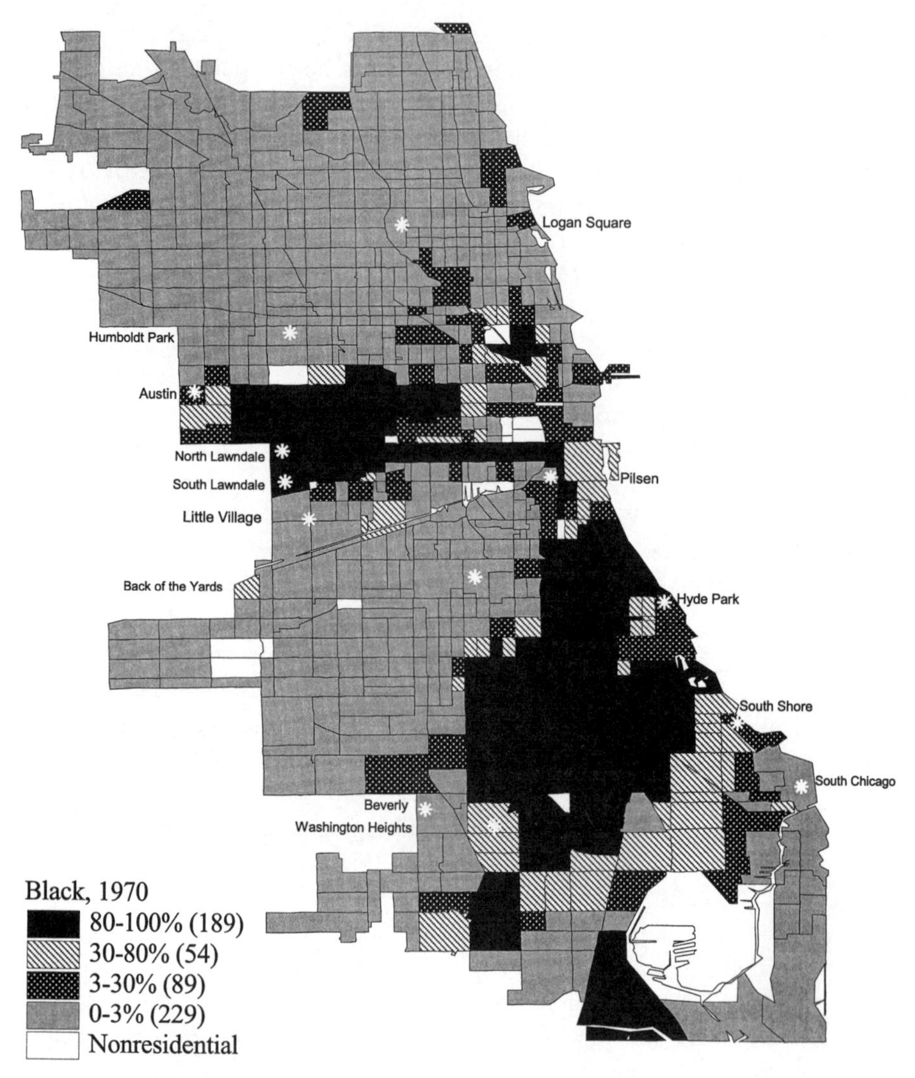

Figure 2.1 Residential Concentration of Black Population, 1970

Although immigration drove the growth of Chicago's Hispanic population, until very recently, no Hispanic neighborhood approached the black ghetto in ethnic homogeneity. In 1970 there were only 31 (of 854) census tracts where at least 50 percent of the residents were Hispanic. Figure 2.3 shows that in 1970 Chicago's Hispanic population was concentrated in four main pockets: (1) a set of northern communities such as Humboldt Park; (2) a cluster toward the center of the city containing the Mexican

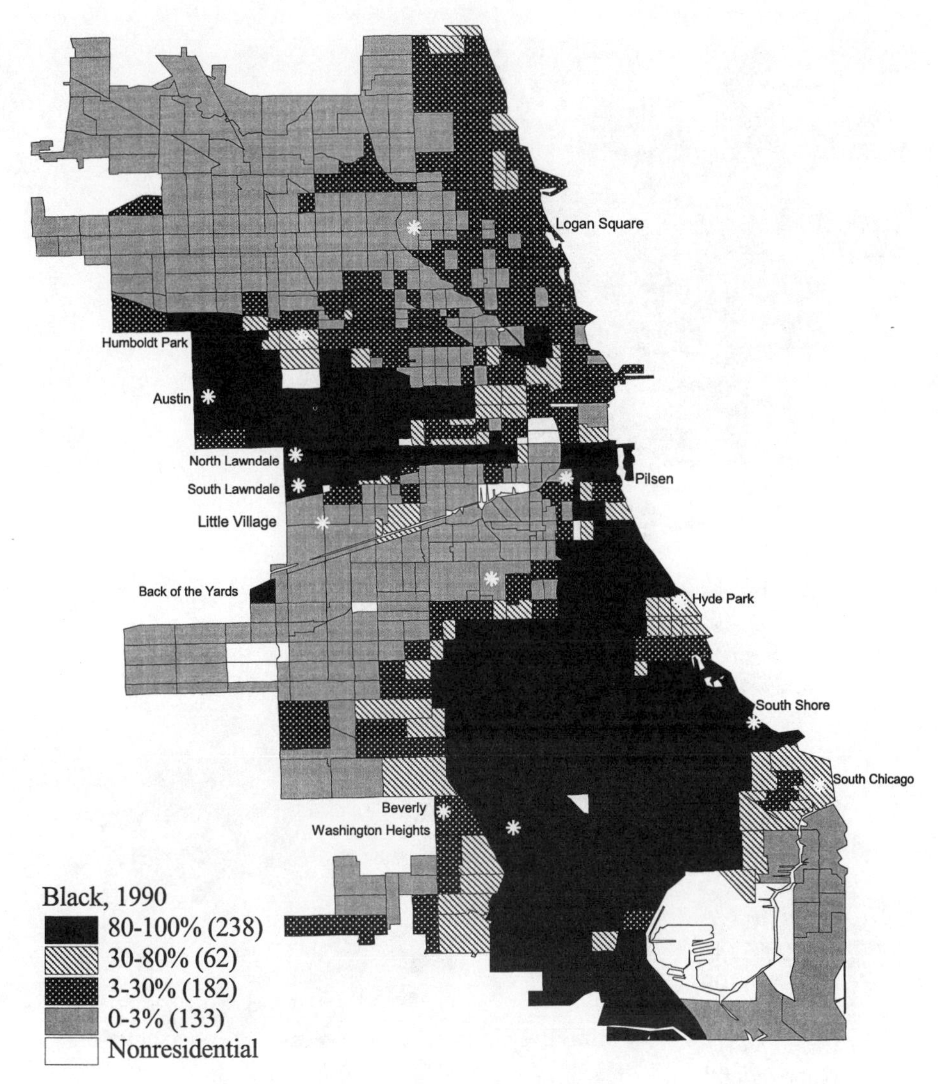

Figure 2.2 Residential Concentration of Black Population, 1990

immigrant communities of Little Village and Pilsen; (3) the near South Side community formerly known as Back of the Yards; and (4) the South Chicago community, which is located in the far southeastern corner of the city. Communities in the northern pocket (such as Humboldt Park, Logan Square, Hermosa, and West Town) are the most heterogeneous of Chicago's Hispanic neighborhoods, each containing sizable proportions of Mexicans, Puerto Ricans, and, in some cases, Cubans.

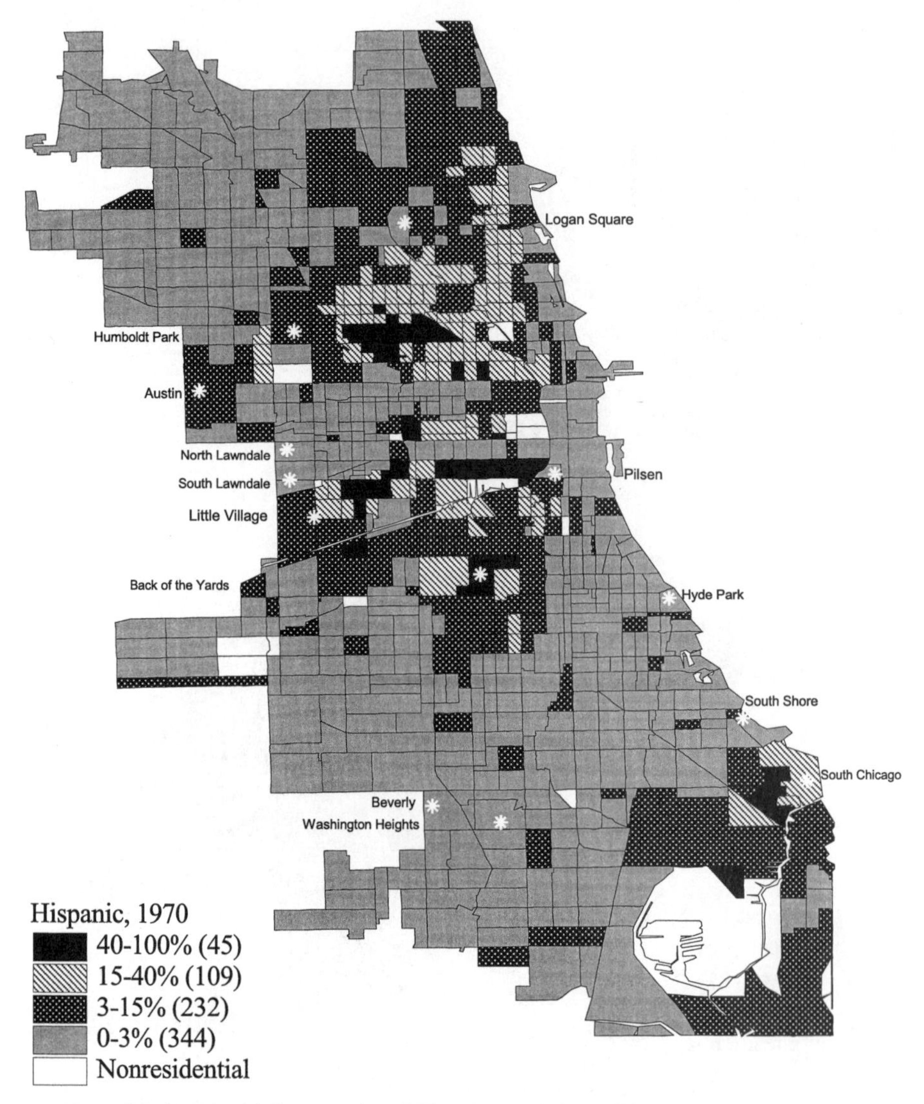

Figure 2.3 Residential Concentration of Hispanic Population, 1970

By 1990 these pockets had expanded into much larger concentrations of Hispanic residents (see figure 2.4), so that 125 census tracts were at least 50 percent Hispanic. This represents a fourfold increase in the number of tracts with a visible Hispanic presence over a two-decade period. But none of these tracts reached a saturation of 100 percent, and only a handful reached 80 percent Hispanic. However, several black groups in

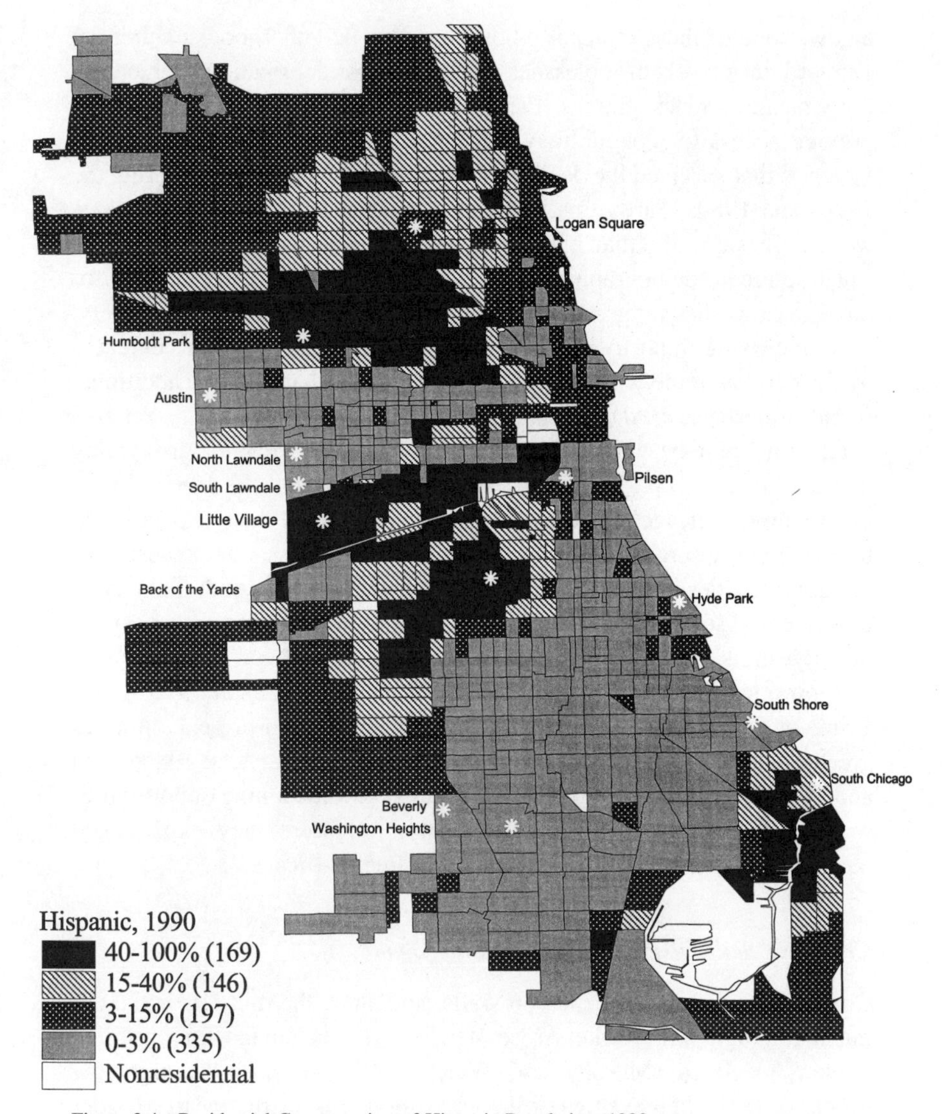

Figure 2.4 Residential Concentration of Hispanic Population, 1990

the highest density census tracts approached complete saturation. The two more centrally located pockets had fused into one large geographical cluster situated in close proximity to older industrial districts near the Chicago River. Finally, the area of Hispanic concentration on the far South Side, near the Lake Calumet industrial district, also expanded to include parts of the South Deering and East Side communities. As we demonstrate below, none of these ethnically homogeneous neighborhoods exhibit the extraordinary levels of joblessness and economic deprivation characterizing Chicago's black ghettos. Partly this is because both Hispanic areas provide access to jobs, as they are situated near the nondurable goods factories that survived the decline of heavy manufacturing jobs during the 1970s and 1980s. Partly this is because Hispanics, especially Mexicans, were more successful than blacks at obtaining jobs in a slack labor market.

These recent demographic trends acquire great social significance for three reasons that suggest how group membership delimits opportunity. First, the rising minority composition of the population was accentuated by the exodus of the white population—a change that acquires additional social importance because it was paralleled by a decline in the relative availability of jobs within the city limits compared to the surrounding metropolitan area (W. J. Wilson 1987; Brune and Camacho 1983). As we show below, Chicago's job base contracted even more than its population base, while the surrounding metropolitan area witnessed an expansion of job opportunities in both absolute and relative terms. Second, high levels of residential segregation further restrict access to job opportunities within the city limits, particularly for groups vulnerable to racial discrimination (Jargowsky 1996; Massey and Denton 1993; W. J. Wilson 1996, 1987; Brune and Camacho 1983). Third, the growing Latino presence implies greater competition for low-wage jobs (Casuso and Camacho 1985; Bean and Tienda 1987; Mare 1995). As we show below, declining opportunities were not uniformly distributed spatially, nor were they experienced equally by whites, blacks, Mexicans, and Puerto Ricans.

Changing Skill Distribution and Labor Demand

Competition for jobs depends on skills, and since the mid-1970s the demand for high-skill workers rose steadily, while demand for unskilled workers declined (Murphy and Welch 1993). Because demographic groups differ in their average skill composition, it is instructive to consider these differences as a way of understanding racial and ethnic economic inequality. Table 2.2, which summarizes the 1970, 1980, and 1990 education distributions of minority and white populations aged twenty-five and over, depicts the structural dilemma confronted by Chicago's black and Latino populations.

To appreciate the extent of racial and ethnic educational inequality, we compare the educational distributions of each group to that of the countywide labor force using the index of dissimilarity (I.D.). The reported

Table 2.2 Educational Attainment by Race and Hispanic Origin: Cook County Residents Ages 25+, 1970–90 (in percent)

	1970	1980	1990
Whites			
<High school	44.4	39	24.6
High school	30.8	30.5	26.4
Some college	12.2	13.2	16.8
College +	12.5	17.3	32.3
N^a	23,869	23,125	31,905
I.D.	4.6	7	5.6
Blacks			
<High school	61.2	46.2	36.2
High school	25.8	29.7	26
Some college	8.7	16.6	21.8
College +	4.2	7.5	16.1
N^a	5,358	14,652	23,944
I.D.	21.4	14.2	15.7
Mexicans			
<High school	81.3	74.8	64.4
High school	12.7	16	18.6
Some college	5.6	6.2	9.6
College +	0.4	3.1	7.5
N^a	482	2,695	6,235
I.D.	41.4	42.8	42.3
Puerto Ricans			
<High school	89.8	68.6	57.6
High school	7.9	20.9	20.1
Some college	1.3	6.7	12.8
College +	1	3.8	9.6
N^a	315	1,113	2,282
I.D.	49.9	37.1	35.5
Total Labor Force[b]			
<High school	39.9	32	22.5
High school	32.3	32	24.4
Some college	14.6	18.5	22.8
College +	13.2	17.5	31.2
N^a	22,649	30,782	63,557

Source: 1970, 1% PUMS; 1980, 5% PUMS; 1990, 5% PUMS

Note: Percentages may not add to 100 due to rounding.

[a] 1990 sample *N*'s unweighted

[b] All employed persons

dissimilarity indices compare the education distributions of black, white, Mexican, and Puerto Rican adults who are not enrolled in school to the education distribution of the total employed population. This comparison serves as a rough indicator of the skill mismatch for each racial-ethnic group.[6] The index values clearly show that throughout the period, Mexicans and Puerto Ricans were more educationally disadvantaged than blacks. A comparison of changes in the index over time indicates a measure of whether the rising educational levels experienced by each group were sufficient to keep pace with the increasing education level of the overall labor force. The disparity between the educational distribution of Cook County's adult black population and the county-wide labor force narrowed over time, from roughly 21 to 16 percent, as did the Puerto Rican educational disadvantage, which fell from 50 to 35 index points. However, the Mexican educational disadvantage remained virtually unchanged over time, due largely to the continued influx of unskilled immigrants during this period. The educational disadvantages among Latinos, and to a lesser extent among blacks, derive from the large shares lacking high school degrees and the tiny shares with postsecondary schooling. Although the share of minority adults with less than a high school education fell appreciably during the period—from 61 to 36 percent for blacks, 81 to 64 percent for Mexicans, and 90 to 58 percent for Puerto Ricans—Chicago's black and Latino populations remained educationally disadvantaged in 1990 compared to whites.

The educational disparities are even more striking when minorities are compared to job holders in the county-wide labor force, where the proportion of adults without a high school degree declined from 40 percent in 1970 to just over 20 percent in 1990. This means that the educational level of the labor force has been rising faster than that of the adult Mexican population. Hence, by 1990, the share of Mexican origin workers lacking a high school education was nearly three times greater than it was among all Cook County job holders. And nearly one-third of the county-wide labor force held college degrees in 1990 compared to 8 percent of Mexicans, less than 10 percent of Puerto Ricans, and 16 percent of blacks. By contrast, one-third of Cook County's 1990 white population held college degrees. This indicates a skill mismatch for minority populations, especially Hispanics, but not for whites.

6. The index of dissimilarity measures the change necessary in the education distribution of each racial-ethnic group to match the education mix of employed workers in Cook County for each year.

Table 2.3 Labor Force Status by Race, Hispanic Origin, and Age: Cook County Men Ages 18–64, 1970–90

	1970		1980		1990	
	LFP	Unemp	LFP	Unemp	LFP	Unemp
Whites, 18–64	86.3	2.2	86.2	6.0	87.4	6.0
18–24	68.9	6.1	79.2	10.6	79.8	9.6
25–44	92.5	1.4	92.6	5.3	93.1	5.7
45–64	87.6	1.6	82.3	4.8	80.4	5.0
N[a]	11,288	10,244	10,195	8,787	13,178	11,522
Blacks, 18–64	75.2	5.9	74.7	16.9	73.5	22.3
18–24	54.4	14.6	61.8	31.5	64.2	37.2
25–44	84.3	4.4	81.8	16.0	79.7	22.9
45–64	75.0	3.8	74.6	8.3	69.1	11.0
N[a]	2,593	2,154	7,212	5,384	10,691	7,855
Mexicans, 18–64	85.7	4.3	89.4	10.2	91.4	9.5
18–24	82.9	—[b]	83.6	15.1	86.1	14.5
25–44	87.4	—	92.3	9.6	95.2	7.5
45–64	84.2	—	90.1	4.9	87.4	8.9
N[a]	328	—	1,920	1,716	4,251	3,887
Puerto Ricans, 18–64	77	4	81	12	78.8	14.7
18–24	65.4	—[b]	68.8	20.3	75.2	19.8
25–44	83.5	—	88.3	9.2	86.4	13.6
45–64	71.4	—	80.8	9.5	66.5	12
N[a]	215	—	739	601	1,303	1,026

Source: 1970, 1% PUMS; 1980, 5% PUMS; 1990, 5% PUMS
[a] 1990 Sample *N*'s are unweighted.
[b] Too small for reliable estimates.

Based on these ethno-racial differences in educational attainment, one would expect that Latinos, and especially Mexicans, would fare worse in the local labor market than blacks. This interpretation assumes that labor market opportunity is driven largely by skill differences, which are proxied by completed schooling. In fact, recent labor force participation trends for Cook County, reported in table 2.3 reveal exactly the opposite, at least for men. Mexican-origin men enjoyed the greatest increases in labor force participation rates, which rose from 86 to 91 percent between 1970 and 1990, while labor force activity rates of black and white men fell slightly or remained stable. In both 1980 and 1990, Mexicans actually had the highest rates of labor force participation of any group. Labor force participation rates include both the employed, that is, persons with jobs, and the unemployed, that is, persons looking for work. Because Mexicans experience higher unemployment rates than whites, the employment rates of Mexicans and whites were similar. The higher unemployment rates of Mexican compared to white men reflects their lower education, which renders them unqualified for many jobs, but their concentration in jobs that are either part-time, temporary, or otherwise unstable also elevates their unemployment rates. Nevertheless, considering their low education levels, Mexicans have unusually high rates of labor force participation, especially compared to blacks.

Age-specific participation and unemployment rates accentuate the racial and ethnic disparities in labor market experiences in several ways. First, the unemployment rates for young black men reached crisis proportions by 1990, when more than one-third of men ages 18–24 who sought work were jobless, compared to 10 percent of whites, 14 percent of Mexicans, and 20 percent of Puerto Ricans. Second, even among men of prime working ages, black unemployment rates were over four times greater than those of comparably aged white men; yet, unemployment rates of Mexican and white men differed by less than 2 percentage points. Third, despite the scenario of declining job opportunities in Chicago, which we elaborate below, labor force participation rates of Mexican-origin men *rose* over time. This paradox also obtains for Puerto Rican men, who, like Mexicans, are educationally disadvantaged compared to blacks. Yet Puerto Ricans fared slightly better than blacks in the labor market on the basis of higher labor force participation rates and lower unemployment rates. On balance, the trends in men's educational attainment and labor force participation suggest that minority-group membership may be more important than educational credentials in determining economic opportunities. Chapter 6 examines this topic in detail.

Women's labor market experiences in Chicago (reported in table 2.4) share some similarities with those of men, but also reveal several notable differences. Unlike men, whose labor force activity rates remained stable or fell slightly (paralleling national trends), participation rates of all women rose appreciably—more so for Latinas than for whites and blacks. The evolution of racial and ethnic differentials in women's labor force activity parallels that of men in three important ways: (1) over time, racial differences widened rather than narrowed, despite converging racial differences in education; (2) gains in Mexican women's labor force participation rates far exceed their educational progress over the period; and (3) unemployment rates are significantly higher for minority women than for their white counterparts. However, there are some noteworthy sex differences as well, most notably women's lower average participation rates with regard to those of men, but also specific ethno-racial differences.

Table 2.4 Labor Force Status by Race, Hispanic Origin, and Age: Cook County Women Ages 18–64, 1970–90

	1970		1980		1990	
	LFP	Unemp	LFP	Unemp	LFP	Unemp
Whites, 18–64	49.4	2.8	65.2	4.9	74.7	4.7
18–24	59.6	3.2	74.0	5.8	77.7	6.4
25–44	44.5	2.9	70.0	4.7	80.6	4.1
45–64	49.8	2.5	56.9	4.6	63.8	5.0
N[a]	12,249	6,468	10,701	6,979	13,334	9,963
Blacks, 18–64	46.7	6.4	57.5	13.7	64.6	16.5
18–24	40.8	12.5	48.3	25.4	55.5	31.4
25–44	50.4	5.5	63.6	12.6	71.0	16.8
45–64	45.1	2.8	54.9	7.6	59.2	7.5
N[a]	3,308	1,728	9,462	5,441	13,659	8,824
Mexicans, 18–64	35.6	6.8	55.3	16.9	62.4	15.1
18–24	47.8	—[b]	62.1	19.9	65.8	18.3
25–44	31.4	—	55.0	16.0	64.5	13.3
45–64	30.8	—	45.9	13.8	51.5	16.4
N[a]	261	—	1,668	923	3,500	2,183
Puerto Ricans, 18–64	30.9	5.6	48.5	10.7	56.2	13.1
18–24	30.2	—[b]	50.6	10.1	57.9	20.8
25–44	37.1	—	51.4	10.7	63.1	11.2
45–64	7.1	—	34.2	11.9	38.3	10.1
N[a]	207	—	794	385	1,394	783

Source: 1970, 1% PUMS; 1980, 5% PUMS; 1990, 5% PUMS
[a] 1990 Sample *N*'s are unweighted.
[b] Too small for reliable estimates.

In 1970, black and white Cook County women participated in the labor force at roughly comparable rates—47 percent and 49 percent, respectively. After 1970, the racial gap in women's average labor force participation widened in Cook County, from 3 percentage points in 1970 to 10 points in 1990. During this period black unemployment nearly trebled, rising from 6 to 17 percent, while white unemployment rose only 2 percentage points. Latinas exhibited the lowest participation rates—36 percent for Mexicans and 31 percent for Puerto Ricans. Ethno-racial disparities in female labor force participation were most pronounced for Puerto Rican women, whose 1990 rate was nearly 20 points below that of white women. Age-specific differences reveal increased activity rates over time for all groups, but also persisting differentials among ethno-racial groups. For Mexican-origin women, the highest labor force activity in 1990 corresponds to women ages 18–24, but for white, black, and Puerto Rican women, participation rates were highest for women ages 25–44. Although this age-specific pattern was common to black women in 1970, it was not characteristic for white women. These patterns suggest that Mexican-origin women withdraw from the labor force when they begin their families, but other groups do not.

Racial and ethnic variation in labor force activity have direct implications for economic well-being, which is manifested through group differences in poverty and welfare participation. We turn to these issues next. In tandem with labor force behavior and family formation patterns, trends and differentials in poverty and welfare participation fueled the underclass debate during the 1980s and early 1990s.

Poverty and Program Participation

The educational and labor market disparities documented above carry over into poverty and welfare participation rates, as shown in table 2.5. For the total population of Cook County, the poverty rate increased by 60 percent during the 1970s, from approximately 12 to 19 percent, but then held relatively steady during the 1980s. Overall trends in means-tested program participation are similar in that they rose steeply during the 1970s, from 4 percent to 12 percent. Not only did rates of poverty and program participation increase during the 1970s, but (as we demonstrate in the next section of this chapter) the geographical concentration

Table 2.5 Poverty and Program Participation by Race and Hispanic Origin: Cook County Families, 1970–90

	1970	1980	1990
Family Poverty			
Whites	8.6	10.4	9.2
Blacks	22.9	30.1	30.9
Mexicans	13.1	21.1	19.3
Puerto Ricans	27.7	33.1	32.6
Total population	11.7	18.9	18.9
Program Participation			
Whites	1.6	3.9	3.2
Blacks	14	22.8	19.6
Mexicans	4	8.1	5.7
Puerto Ricans	16.4	23.2	22.6
Total population	4.2	11.6	10.14
N[a]	17,644	26,231	36,797

Source: 1970, 1% PUMS; 1980, 5% PUMS; 1990, 5% PUMS.
[a] 1990 sample N's unweighted.

of poverty, together with other aspects of socioeconomic disadvantage, grew exponentially from 1970 to 1980 and then remained high, albeit stable, during the 1980s.

A second important observation is that blacks and Puerto Ricans fare much worse than whites in terms of poverty rates and welfare participation, while Mexicans occupy the middle ground. These differentials parallel those in labor force activity. Cook County's Puerto Rican population was the most economically disadvantaged group, as is evident both in poverty rates and participation in means-tested programs. Over time, poverty differentials increased between whites and other groups, so that by 1990, black and Puerto Rican families were approximately three-and-a-half times more likely to be poor than white families; Mexican families were twice as likely to be poor when compared to whites. Thus, unlike the group differences in education, which converged over time, the gap between minority and white poverty rates rose for all groups, but especially for blacks and Puerto Ricans.

Like trends in labor force participation, these poverty differentials, together with the labor market outcomes presented above, underscore the significance of color, as socially constructed, in circumscribing economic opportunity. Educational achievement may be a necessary condition for accessing labor market opportunities, but it is obviously insufficient to explain the inferior labor market position and economic status of blacks relative to Latinos in Chicago or elsewhere. However, thus far we have ignored two important issues that bear on the color and opportunity theme. One is the trend in job growth and its spatial distribution relative to changing population growth, composition, and distribution. The other concerns the nature and severity of residential segregation by race and Hispanic origin, for this ultimately delimits economic opportunities.

Accordingly, the remainder of this chapter is devoted to these issues. The following section traces changes in employment over time and space, and relates these to the ethno-racial configuration of Chicago's population. We show that, in contrast to the nearly two-decade period following World War II, the early 1970s through the mid 1980s was a period of slow economic growth accompanied by substantial changes in the industrial composition of employment and by stagnating real wages for Chicago's industrial workers.

Spatial and Temporal Changes in Opportunity

Like most northern industrial cities, Chicago began a process of industrial restructuring that resulted in the loss of jobs and a recomposition of the industrial structure of employment during the 1970s. For the entire metropolitan area, job losses must be understood in spatial terms, because declining employment within the city limits was accompanied by impressive job growth in the surrounding suburbs. Furthermore, the changing spatial configuration of employment was accompanied by shifts in its industrial composition away from relatively well-compensated factory jobs toward relatively low-wage service jobs. These structural changes in the spatial distribution and industrial composition of employment altered economic opportunity along racial and ethnic lines because minority populations became a larger proportion of the central-city population just when job growth within the city limits stagnated. This dilemma—dubbed the spatial "mismatch hypothesis" (Holzer 1991)—captures a fundamental disjuncture between *where workers work* and *where they live.* Therefore, to describe the changes in the ecological transformation of employment, we first compare changes in job growth (and decline) in the six-county SMSA surrounding Chicago, in Cook County (which includes the City of Chicago and its immediate surrounding suburbs) and in the city proper.

Although precise estimates of job losses vary by source, there is general agreement that the City of Chicago lost between a hundred and fifty thousand and two hundred thousand jobs during the 1970s and 1980s—approximately 10–15 percent of the 1970 job base. Table 2.6, which presents changes in Unemployment Insurance (UI) employment for the City of Chicago, Cook County, and the six-county SMSA, reveals three distinct trends. First, and most important, the City of Chicago experienced a net job loss from 1972 to 1990, while manufacturing and nonmanufacturing employment grew by 10 percent in Cook County and a vigorous 27 percent in the six-county metropolitan area.[7] These opposed trends portray a major change in the ecology of job opportunities within and beyond the city limits. If, as several studies suggest, place of residence delimits access to nearby jobs, then the increased concentration of blacks and Hispanics *within* the city limits imply greatly reduced economic opportunities for Chicago's minority workers.

Second, the rate of job loss and growth reflects both structural and cyclical changes. The aggregate changes between 1970 (1972) and 1990 did not occur uniformly but fluctuated in accordance with the business cycle. The economic peaks and troughs are most clearly evident in the

7. Although we present the data series from 1970 to 1990, the data for 1970 and 1971 are not strictly comparable with the series beginning after 1972. Therefore, we base our substantive discussion on the comparable segment of the series.

Table 2.6 Changes in UI-Covered Manufacturing and Nonmanufacturing Employment: Chicago Metro Area, 1970–90 (Thousands)

	Six-County SMSA		Cook County		City of Chicago	
Year	Manuf	Nonmanuf	Manuf	Nonmanuf	Manuf	Nonmanuf
1970	948	1,337	802	1,166	497	821
1971	867	1,326	733	1,145	439	795
1972	868	1,545	729	1,315	435	888
1973	n/a	n/a	n/a	n/a	440	877
1974	917	1,658	761	1,377	441	893
1975	801	1,612	656	1,350	376	864
1976	794	1,657	650	1,371	367	858
1977	769	1,616	625	1,330	350	824
1978	789	1,734	634	1,412	352	854
1979	805	1,795	646	1,462	354	878
1980	n/a	n/a	n/a	n/a	n/a	n/a
1981	744	1,832	593	1,483	319	881
1982	698	1,778	557	1,439	296	850
1983	634	1,817	499	1,445	258	844
1984	674	1,895	520	1,495	264	862
1985	n/a	n/a	510	1,549	257	880
1986	628	2,027	478	1,574	239	890
1987	n/a	n/a	n/a	n/a	n/a	n/a
1988	648	2,247	482	1,710	237	973
1989	656	2,335	474	1,758	225	987
1990	653	2,402	470	1,774	216	984
% Δ1972–1990	−24.8	55.5	−35.5	34.9	−50.3	10.8
1972–1979	−7.3	16.2	−11.4	11.2	−18.6	1.1
1981–1990	−12.2	31.1	−20.7	19.6	−32.3	11.7

Source: Illinois Department of Employment Security, *Where Workers Work in the Chicago Metro Area,* Chicago: Economic Information and Analysis Division, Selected Years
Note: Numbers rounded to nearest thousandth. No reports were issued for 1980 and 1987.

nonmanufacturing employment trend, which generally increased throughout the period except during recessions. Furthermore, the pace of economic recovery was not spatially uniform. This is apparent in the sluggish nonmanufacturing job growth in the city compared to the surrounding suburbs and adjacent counties during the 1970s and postrecession 1980s.

Third, nonmanufacturing jobs increased appreciably over the two-decade period, while manufacturing jobs declined as a share of total employment. The industrial composition of employment changed so that between 79 percent (Cook and SMSA) and 82 percent (City of Chicago) of all jobs in 1990 involved nonmanufacturing concerns, primarily services. The loss of manufacturing jobs that accompanied the recession of the early 1970s was accentuated once again during the economic downturn at the beginning of the 1980s. Specifically, between 1970 and 1979 the city lost more than 143,000 manufacturing jobs at a time when nonmanufacturing jobs increased by a meager 57,000. The city lost an additional 103,000 manufacturing jobs between 1981 and 1990, which was fully compensated for by an increase of roughly the same number of nonmanufacturing jobs. For Cook County, the loss of manufacturing jobs from 1970 to 1990 exceeded 330,000, yet the county-wide growth of nonmanufacturing employment not only offset the net loss of manufacturing jobs but nearly doubled it (608,000). Although Cook County's manufacturing sector witnessed a modest rebound during the mid-1980s, this was neither sustained nor paralleled in the city, which continued to lose manufacturing jobs even as the nonmanufacturing sector recovered after 1982.

These uneven trends in job growth and industrial transformation are portrayed graphically in figures 2.5 and 2.6. Rather than comparing the City of Chicago with all of Cook County and the entire six-county SMSA, these graphs compare trends in job growth across (1) the City of Chicago, (2) the *non-Chicago* portion of Cook County, and (3) the *non–Cook County* portion of the SMSA. Consequently, figures 2.5 and 2.6 sharpen the geographical comparisons by showing how changes in employment within the city limits compared to those in the surrounding suburbs as well as the more geographically remote suburbs.

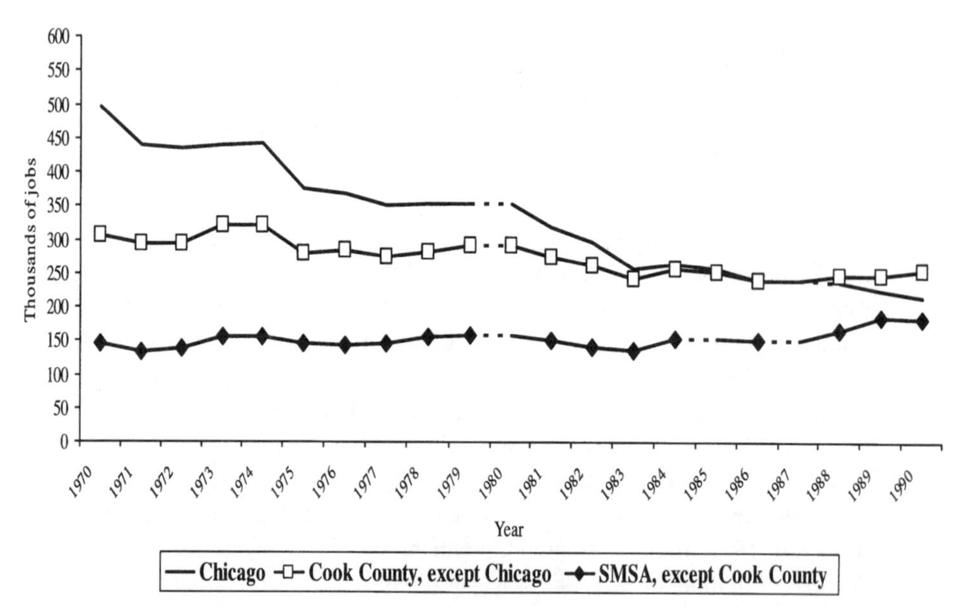

Figure 2.5 Changes in UI-Covered Manufacturing Employment by Place: Chicago Metro Area, 1970–90

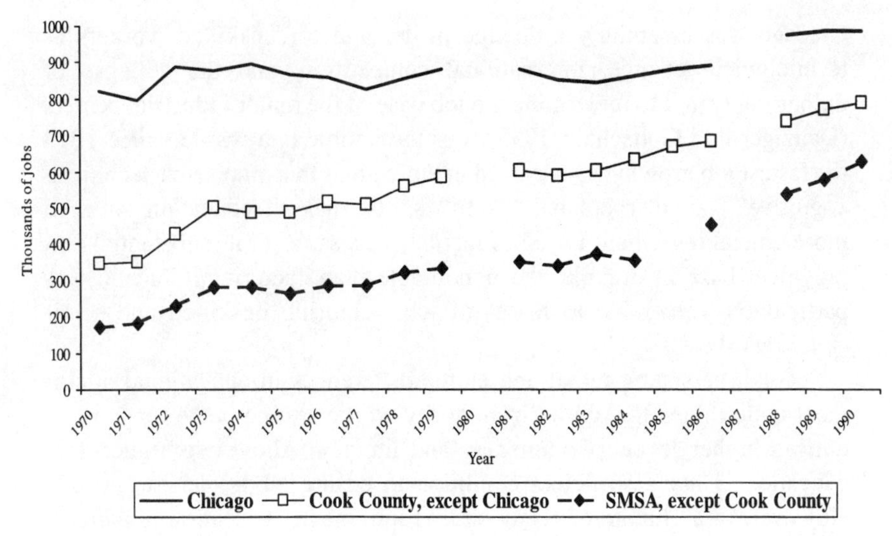

Figure 2.6 Changes in UI-Covered Nonmanufacturing Employment by Place: Chicago Metro Area, 1970–90

Figure 2.5 shows changes over time in manufacturing employment over the three non-overlapping areas. Whereas the outlying portions of Cook County and the SMSA held a relatively steady share of manufacturing jobs over time, the City of Chicago lost more than 56 percent of its manufacturing base between 1970 and 1990. Over the same period, the outlying portion of Cook County lost only 17 percent of its manufacturing base, while the non–Cook County portions of the SMSA registered a 25 percent increase in manufacturing jobs. This means not only that the industrial decline was disproportionately concentrated in the city, but also that most of the new manufacturing jobs arose in the most geographically remote counties in the SMSA. All three ecological areas experienced a growth in nonmanufacturing employment, as displayed in figure 2.6, but the growth rate was much higher in the outlying suburbs than within the city limits. From 1970 to 1990, the outlying counties of the SMSA experienced a staggering 267 percent increase in nonmanufacturing jobs, compared to increases of 128 percent in the portion of Cook County outside the city limits, and a meager 20 percent in Chicago.

These changes in the industrial composition of employment are profound in three major ways that relate to our theme of color and opportunity. One is that the shift away from manufacturing and toward service employment fostered changes in both the skill requirements of jobs and the compensation provided. The decline in manufacturing employment in

Chicago was essentially a decline in demand for unskilled workers as technological change, international competition, and the collapse of unions conspired to undermine the job base of the major industrial centers (Danziger and Gottschalk 1993). For the nation as a whole, since 1973 the fastest job growth has occurred in those areas that emphasize technical, cognitive, and interpersonal ("soft") skills, making education an even more crucial ingredient for labor market success. With its substantial employment base in durable and nondurable manufacturing, Chicago was particularly vulnerable to losses of jobs requiring unskilled and semiskilled workers.

Second, persisting racial and ethnic differences in educational attainment undermined the ability of minority workers to compete for jobs requiring higher levels of numeracy and literacy. Above we argued that education is not a sufficient condition to ensure labor market success. However, in a climate of rising skill requirements, education is increasingly necessary to access the best-compensated jobs. Also, in slack labor markets, educational attainment is used as a screen to identify more capable, if not more productive, workers. Finally, job losses and gains *within* the city limits were unevenly distributed, and this further accentuated the negative impact of industrial restructuring on minority populations which, as shown above, are disproportionately concentrated within the city limits relative to the suburban county fringe. Moreover, within the city limits, job losses were greatest in neighborhoods beyond the greater "Loop" and its surrounding business fringe, while net job gains accrued to surrounding suburbs (PIChicago 1995).

For example, PIChicago estimates that between 1974 and 1983, when employment fell 17 percent citywide, the decline in the greater Loop area was under 5 percent compared to 24 percent in other neighborhoods. During this period, employment rose by 9 percent in the metro area suburbs. From 1984 to 1990, employment rose by nearly 12 percent in the greater Loop area, by 30 percent in the surrounding metro areas, and fell by 0.5 percent in the remaining inner-city neighborhoods. The recession of the early 1990s once again took a differential toll on employment opportunities within the city limits. Employment declined by 4 percent in the Loop between 1991 and 1994, but it fell 6 percent in the remaining inner-city neighborhoods. Suburban employment opportunities rose 7.7 percent during this period (PIChicago 1995, 5).

The significance of the uneven geographic distribution of job growth and decline is that even during periods of brisk economic recovery, the associated improvement in job opportunities remains spatially and temporally circumscribed. As we show below, this change in the spatial ecology of jobs had profound implications for the social ecology of urban life and the economic well-being of minority populations in particular. A second sobering lesson is that the citywide job gains associated with the recovery of the late 1980s were virtually dissipated by the economic downturn of the early 1990s. Another important lesson from the trends in job growth and decline is that both structural and cyclical changes drove shifts in the spatial distribution of employment—*where workers work*—and that these changes also transformed in profound ways the social fabric of urban neighborhoods—that is, *where workers live.* The following section traces the social consequences of spatially delineated job opportunities for the two-decade period under consideration.

Neighborhoods in Transition

Concurrently, and perhaps fueled by a massive economic restructuring in Chicago during the 1970s and 1980s, the city also witnessed a profound social transformation that altered the character of its neighborhoods. One part of this story, which has received enormous attention from social scientists during the 1980s, involves the increasing concentration of poverty in predominantly black ghetto neighborhoods. Most scholars demonstrated that as a set of institutional arrangements, the ghetto confines, isolates, and excludes; most important, however, it limits opportunity (Massey and Denton 1993; W. J. Wilson 1996, 1987). Another part of the story, which has received much less attention, concerns changes in the overall social ecology that accompany spatially concentrated poverty.

Although indisputable evidence exists that urban inequality increased during the 1970s and 1980s, as evidenced by sharp increases in the concentration of various measures of economic deprivation and social disadvantage (W. J. Wilson 1987; Jargowsky 1996), surprisingly few attempts have been made to situate this trend toward concentrated poverty within the broader context of urban social change. If concentrated urban poverty is coupled with the demise of working-class neighborhoods and a growing concentration of affluence, the social implications of spatially distributed economic opportunities transcend the boundaries of high-poverty neighborhoods. Equally important, the spatial polarization of neighborhoods is crucial for understanding how group membership circumscribes opportunity in Chicago.

The remainder of this chapter redresses this shortcoming in prior research by examining changes in Chicago's ecological structure, analyzing the patterns of neighborhood change that undergird these transformations, and clarifying how ecological change differentially restricts spatially distributed economic opportunity along racial and ethnic lines. The term *ecological* signifies that our analytic units—neighborhoods—are spatially defined aggregates that are proxied by census tracts. Hence, *ecological structure* refers to the stratification of neighborhoods over time and space. Likewise, *ecological characteristics* refer to neighborhood-level measurements. For example, many sociodemographic characteristics, such as poverty, unemployment, and race, reflect categorical distinctions at the individual level but are typically expressed as percentages or rates when translated into ecological constructs.

We first develop an empirical typology of Chicago neighborhoods that classifies neighborhoods based on a multidimensional array of characteristics, including socioeconomic status, age composition, and residential stability. This typology allows us to examine the ecological structure of Chicago neighborhoods and to depict how the paths of neighborhood change have transformed that structure over time. Relying on this typology, we address three significant yet neglected questions about the socioeconomic implications of trends in neighborhood inequality in Chicago. The first concerns changes in the distribution of different neighborhood types (e.g., middle-class, working-class, and underclass) throughout the city. What types of neighborhoods have become more or less prominent in the city's ecological structure over time? Although the growth of underclass or ghetto-poverty neighborhoods has been well documented (e.g., Jargowsky 1996; Jargowsky and Bane 1991), has there been a countervailing trend toward the concentration of affluence that would result in a more polarized ecological distribution? The second set of questions pertains to the paths of neighborhood change. What types of neighborhoods were most vulnerable to extreme economic decline? Which ones gentrified or otherwise upgraded, and which remained ecologically stable over time? Finally, who gains and who loses from these changes in urban structure? For instance, are black middle- and working-class areas more or less likely to maintain stability than similar white areas? Where do emerging Hispanic populations fit into the ecological structure, and what types of neighborhoods are they likely to inhabit? Answers to these questions provide important insights about how economic opportunity is circumscribed along racial and ethnic lines.

Relying on tract-level census data for the City of Chicago from 1970 to 1990, our empirical strategy entailed constructing a typology of Chicago neighborhoods that categorizes census tracts into ecological categories at each of the three time periods. This framework allows us to view transformations in the overall ecological structure of Chicago as certain categories grow or decline over time, and to analyze the paths of neighborhood change as individual neighborhoods move in and out of different categories over time. Moreover, our methodological approach overcomes several deficiencies in the portrayal of neighborhood change in most studies of urban poverty. The conventional approach is to classify neighborhoods into different categories on the basis of arbitrary cutoff points in the distribution of a single variable, usually individual or family poverty rates.[8]

We found that this approach had three drawbacks. First, most studies employing this approach focus on neighborhoods situated at one tail of the distribution, namely, those with the highest poverty rates, revealing little about potentially important distinctions among the group of so-called nonpoor neighborhoods. Second, the cutoff points are usually imposed arbitrarily. Finally, as we demonstrate below, single-indicator approaches neglect other important sociodemographic dimensions of a city's ecological structure, such as age composition and residential stability, which are often decisive in shaping trajectories of neighborhood change and the attendant opportunities for population groups.

Therefore, we constructed ecological categories across a multidimensional array of empirically observed neighborhood characteristics, including measures of socioeconomic status, residential stability, and family structure. More specifically, we conducted a cluster analysis on 825 Chicago census tracts (collected for the census years 1970, 1980, and 1990), with a set of ten variables. These include rates of poverty and public assistance use, percentage of college graduates, percentage of white-collar workers, percentage of female-headed families, owner-occupancy rates, residential stability (if residents lived in the same house for at least five years), and two indicators of age structure (percent residents aged seventeen and younger, and percent aged seventy-five and older). Cluster analysis is a statistical technique that groups cases (in our case census tracts) into hierarchical categories on the basis of their "proximity" to one another.[9]

8. As an example, neighborhoods with a poverty rate greater than 40 percent are often defined as "ghetto poor"; those with rates between 20 and 40 percent may be labeled simply as "poor"; and those with poverty rates under 20 percent are usually lumped into the category of "nonpoor." Changes in the concentration of poverty are then assessed by the number of neighborhoods that enter these categories over time. See Ruggles (1990) for further discussion of alternative poverty measures.

9. We relied on Ward's minimum-variance algorithm to compute the similarity measures (see SAS 1990).

Social Ecology of Chicago Neighborhoods

Our cluster analysis yields a fourfold typology of Chicago neighborhoods from 1970 to 1990 that consists of the following ecological categories: (1) stable middle-class neighborhoods; (2) gentrifying "yuppie" neighborhoods; (3) transitional working-class neighborhoods; and (4) ghetto underclass neighborhoods.[10] The clustering procedure itself provides no insights into the meaning of the categories it creates; it simply classifies neighborhoods on the basis of their similarities. Our labels for the four clusters reflect our interpretation of neighborhood types based on the salience of various characteristics and our firsthand knowledge of the city. To arrive at our interpretations, we examined descriptive statistics for each of the clusters, as presented in table 2.7.[11] Although measures of racial and ethnic composition were not used in the cluster analysis, their mean values are also displayed in order to show how the resulting ecological types vary with respect to minority population. As a baseline reference for comparing clusters, the first column of table 2.7 displays means and standard deviations for the entire city.[12]

The first cluster consists mainly of residentially *stable middle-class neighborhoods* with an aging population. Neighborhoods so classified tend to be of high socioeconomic status, as indicated by the low rates of poverty, public assistance, unemployment, and female-headed families relative to the citywide means. The only factors that detract from the high socioeconomic status of these neighborhoods are the moderate levels of both college graduates and, to a lesser extent, white-collar workers. These neighborhoods contain high percentages of elderly and small to moderate-sized youth populations. Mean rates of owner occupancy and residential stability are both very high in this category. Stable middle-class neighborhoods contained relatively few blacks and Hispanics, although these levels did increase over time. For example, whereas in 1970 the average tract in this group was only 7.3 percent black, by 1990 the average tract was

10. Individual neighborhoods can move through different ecological categories over time because the unit of analysis is not the tract per se, but rather the tract-year. As a result, the categories themselves can grow or decline in size over time based on the number of neighborhoods they include.

11. We also examined how the characteristics of each cluster changed over time, but in the interest of parsimony we do not present these tables.

12. The citywide tabulations do not represent the true mean values for the entire city because they are calculated to reflect average characteristics of census tracts from different time periods and with varying population sizes. Thus these means are subject to aggregation bias.

Table 2.7 Descriptive Statistics by Neighborhood Type

	Citywide		Stable Middle-Class		Gentrifying Yuppie		Transitional Working-Class		Underclass	
Variable	Mean	(Std Dev)	Mean	(Std Dev)	Mean	(Std Dev)	Mean	(Std Dev)	Mean	(Std Dev)
Used in Clustering										
%Poor	18.1	(16.9)	7.2	(6.3)	8.8	(7.9)	20.7	(10.4)	43.3	(16.6)
%Public assistance	14.7	(15.9)	6.1	(6.5)	5.6	(5.8)	13.4	(8.5)	41.4	(14.3)
Unemployment rate	10.5	(9.2)	6.7	(4.7)	5.2	(3.7)	9.1	(4.9)	24.6	(10.5)
%College graduate	15.9	(15.9)	13.7	(8.6)	48.8	(17.7)	10.0	(7.5)	6.4	(4.9)
%White collar	55.3	(18.9)	56.8	(13.4)	79.3	(12.6)	42.7	(12.6)	53.4	(21.9)
%Female head	23.7	(17.8)	14.0	(9.2)	17.7	(12.4)	22.1	(10.1)	50.5	(16.9)
%17 and under	29.1	(11.1)	25.2	(6.4)	14.8	(7.4)	35.6	(7.8)	38.0	(10.1)
%75 and older	11.1	(6.4)	14.4	(6.3)	12.2	(6.6)	7.9	(3.9)	8.4	(5.6)
%Same house 5 yr	55.1	(15.0)	63.9	(10.2)	38.2	(11.0)	48.0	(13.3)	61.1	(11.2)
%Owner occupancy	36.1	(24.4)	56.7	(22.0)	21.0	(14.5)	25.1	(14.9)	22.5	(16.6)
Not used in clustering										
%Black	37.5	(43.9)	17.8	(35.9)	17.6	(24.5)	39.3	(42.9)	90.1	(22.9)
%Hispanic	14.8	(22.2)	9.9	(15.7)	10.2	(10.8)	29.1	(29.0)	6.5	(17.5)
%Foreign-born	10.3	(13.1)	9.7	(11.2)	13.4	(11.6)	14.5	(16.5)	2.7	(7.0)

Source: U.S. Census 1970–90, City of Chicago

26.3 percent black. Thus, despite changes in its racial composition, this cluster maintained fairly stable socioeconomic characteristics over time.[13]

Gentrifying yuppie neighborhoods are also characterized by relatively high socioeconomic status. However, these neighborhoods differ from stable middle-class neighborhoods in their distinctive age structures and unstable residential character. Youth are relatively underrepresented in these areas, and although the percentage of elderly is slightly higher than the citywide average, auxiliary tabulations revealed that the elderly population declined over time. Another distinguishing feature of this cluster is its low level of residential stability. Less than 40 percent of persons residing in gentrifying yuppie neighborhoods lived in the same house for at least five years, and the mean owner occupancy rate was only 21 percent. Rates of poverty, public assistance, unemployment, and female headship are low in these neighborhoods, while the share of college-educated residents and white-collar workers is higher than in stable, middle-class areas of the city. Whites predominate in these neighborhoods.

The third cluster, *transitional working-class neighborhoods,* plays a pivotal role in the process of neighborhood change. Compared to the gentrifying yuppie and middle-class neighborhood types, these areas exhibit lower socioeconomic status but are close to citywide averages along many dimensions. For example, the poverty rate in these neighborhoods is just above the citywide mean, while rates of public assistance, unemployment, and mother-only families fall just below the citywide averages. However, the most distinctive socioeconomic characteristics of these neighborhoods are their low shares of college-educated residents and white-collar workers. Demographically, these neighborhoods are characterized by high percentages of youths and low percentages of elderly. Levels of residential stability and owner occupancy also fall below the citywide averages, which underscores the transitional nature of these neighborhoods. Perhaps the most distinguishing characteristic of transitional working-class neighborhoods is their racial and ethnic composition, which became increasingly dominated by Hispanics and immigrants over time. Unreported tabulations show that the Hispanic presence in this cluster increased from 14 to 58 percent of the population between 1970 and 1990.

13. An important caveat is that stability along the socioeconomic dimension is a direct result of the clustering procedure. This does not mean that individual neighborhoods in this cluster necessarily remained economically stable. Rather, the cluster retained stable characteristics, which is a direct outcome of the clustering procedure. As we demonstrate below, this cluster comprises a different set of neighborhoods at each time point.

The final neighborhood type comprises socially and economically disadvantaged neighborhoods, which have been dubbed *ghetto underclass neighborhoods,* or ghetto-poverty areas, which suffer from severe socioeconomic deprivation. Rates of poverty, unemployment, public assistance, and female-headed families are very high over time, while the proportion of college graduates is very low. Of the socioeconomic characteristics considered, only the proportion of white-collar workers approaches the city average. The structural context of disadvantage in this cluster is further compounded by high rates of residential entrenchment (the percentage of people living in the same house for at least five years) and low rates of owner occupancy. Like transitional working-class neighborhoods, these areas are characterized by a high proportion of youths and a low proportion of elderly residents. Another distinctive characteristic of these areas is their racial homogeneity, which averaged 90 percent black, although Hispanics increased their presence in this category over time from 1 percent in 1970 to 9 percent in 1990.[14] The social significance of these neighborhood types hinges on their relative prominence and spatial configuration in the social ecology of the city.

Geographical Distribution of Neighborhood Types

The 1970, 1980, and 1990 neighborhood typology is mapped in figures 2.7, 2.8, and 2.9, respectively. In 1970, stable middle-class neighborhoods were located in the western portions of the city on both the North Side in predominantly white communities (such as Edison Park, Norwood Park, Jefferson Park, and Forest Glen), and on the South Side in both white areas (such as Clearing, West Lawn, Ashburn, and Mount Greenwood) and relatively prosperous black communities (notably Beverly and Washington Heights). This map also shows that the stable middle-class and gentrifying yuppie categories, which share many similar socioeconomic characteristics, occupy different geographic spaces. Gentrifying yuppie neighborhoods are located almost exclusively on the eastern edge of Chicago's lakefront, in communities such as Lincoln Park, Lakeview, and Rogers Park on the North Side; and on the South Side in Douglas, Hyde Park (home of the University of Chicago), and South Shore.

Transitional working-class neighborhoods were the dominant ecological category in 1970, encompassing many of the interior portions of the

14. In part, this reflects the use of census tracts as analytic units and the spatial contiguity of the black ghetto and the burgeoning Mexican immigrant neighborhood in the city.

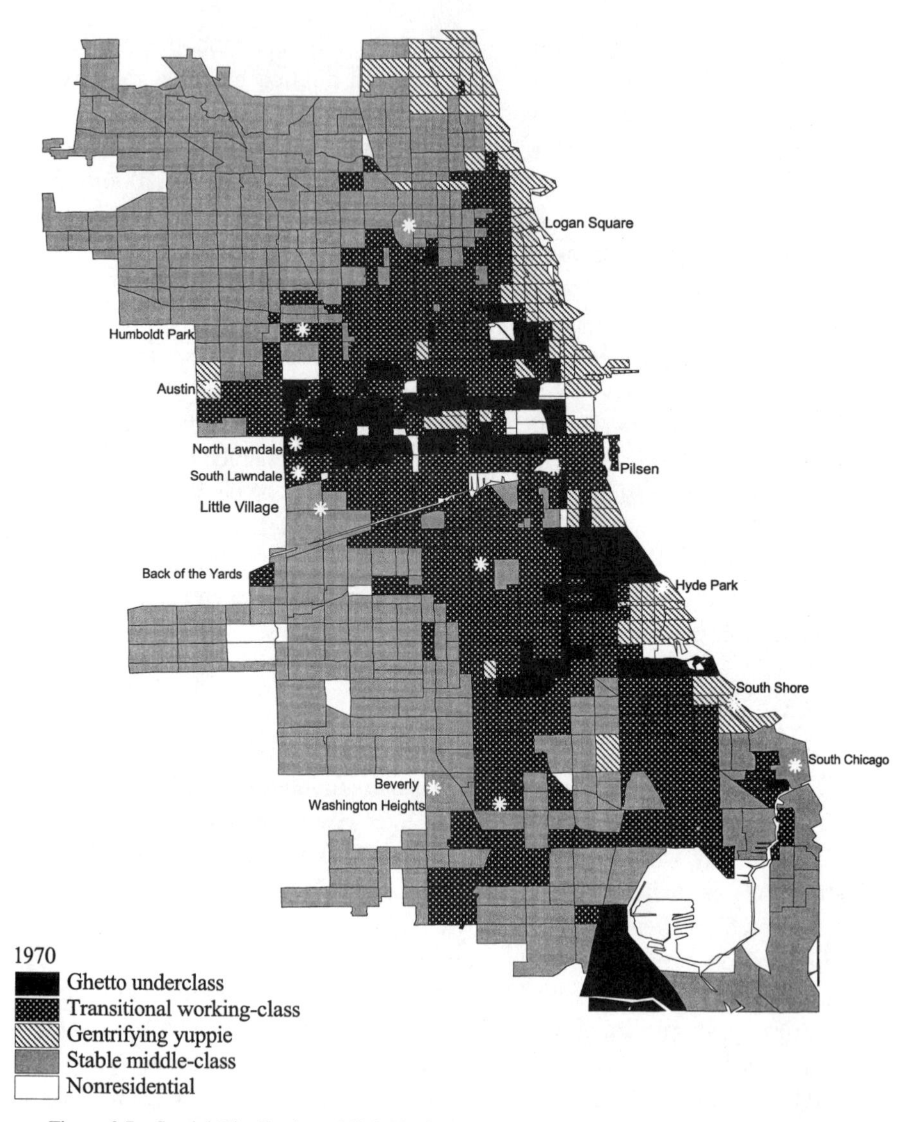

Figure 2.7 Spatial Distribution of Neighborhood Types, 1970

city, including several emerging Hispanic neighborhoods, such as Logan Square, Humboldt Park, West Town, McKinley Park, South Lawndale, and South Chicago. Working-class neighborhoods also embraced some black communities on the South Side, such as West Englewood and Auburn Gresham, in 1970. Ghetto underclass neighborhoods are almost entirely confined to the core of the Chicago's traditional black belt—in areas

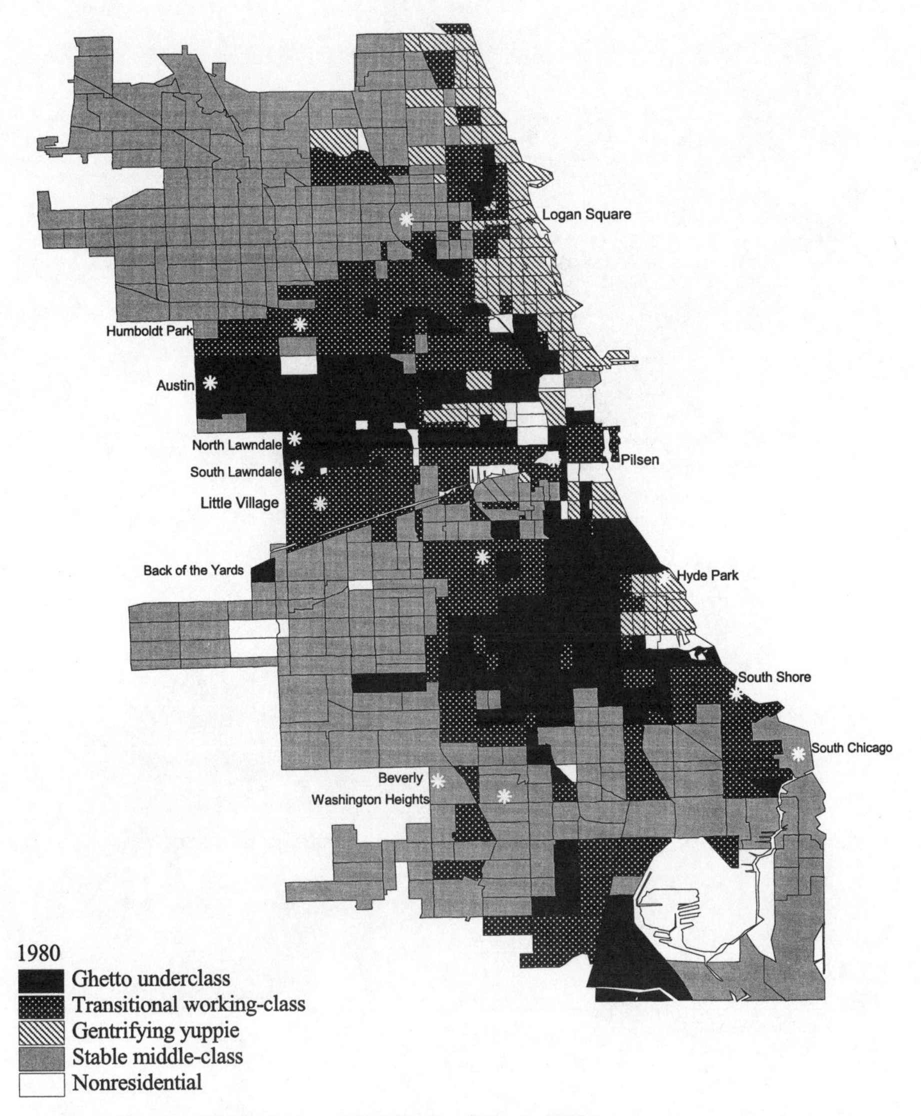

Figure 2.8 Spatial Distribution of Neighborhood Types, 1980

such as Oakland, Grand Boulevard, and Englewood to the south, and North Lawndale to the west.

Figure 2.8 reveals that the most dramatic change in Chicago's ecological landscape is the expansion of ghetto underclass neighborhoods along the peripheries of both the southern and western corridors of the black belt. The rapid economic transformation presents as ghetto expansion,

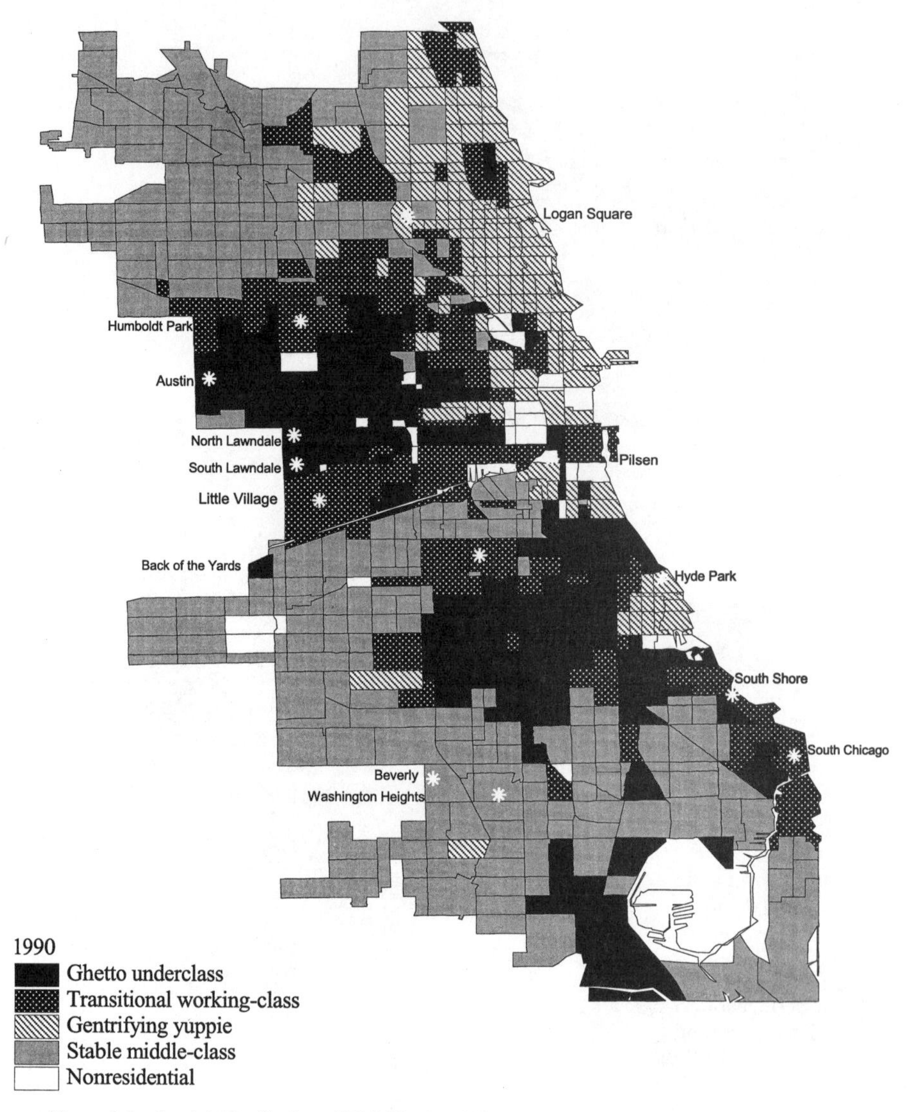

Figure 2.9 Spatial Distribution of Neighborhood Types, 1990

which occurred primarily at the expense of former working-class communities such as Austin, West Garfield Park, and North Lawndale to the west and West Englewood, Washington Heights, and Roseland to the south. However, other transitional working-class neighborhoods that were on the periphery of the southern corridor of the black belt, such as Avalon Park, Calumet Heights, Pullman, and South Deering upgraded and entered the stable middle-class category. A substantial number of black neighborhoods on the far South Side experienced upgrading. This suggests that the process of ghetto expansion is not inevitable for neighborhoods caught in the wake of segregation, although it is highly likely because of segregation.

Whereas the dramatic expansion of the ghetto was the most notable change in Chicago's socioeconomic and ethno-racial landscape during the 1970s, the principal story of the 1980s, as shown in figure 2.9, is the spread of gentrification along the lakefront at the eastern edge of the city. However, the impact of gentrification on Chicago's ecological landscape was small by comparison to ghetto expansion, which continued during the 1980s and began to incorporate Puerto Rican areas in northwest parts of the city, including sections of Humboldt Park and Logan Square. At the same time, however, and despite the spatial encroachment of the ghetto, many black middle-class neighborhoods remained ecologically stable over time. In particular, those black neighborhoods that upgraded to the stable middle-class category during the 1970s tended to remain in that category during the 1980s.

Paths of Neighborhood Change

To examine the growth and decline of the ecological categories, we cross-classified the four types of neighborhoods over successive time periods. Like mobility tables, the cross-tabulations displayed in table 2.8 depict intertemporal "inflows" and "outflows" among ecological categories. The top panel portrays patterns of change during the 1970s, the middle panel shows mobility during the 1980s, and the bottom panel displays neighborhood trajectories over the two-decade period. These tables reveal that both the stable middle-class and transitional working-class categories diminished in size over time. The transitional working-class category was the most dominant neighborhood type in 1970, comprising 45 percent of all tracts (374 out of 825), but it diminished considerably over each decade and accounted for only 14 percent of all neighborhoods (tracts) in 1990 (116 out of 825).[15] The stable middle-class category also experienced a decline, albeit less drastic, shrinking from 41 percent (342 out of 825) of all tracts in 1970 to 34 percent (278 out of 825) in 1990.

Categories at the two poles of the socioeconomic distribution—the

15. Although Chicago has 854 census tracts, this analysis is restricted to the 825 tracts with constant boundaries over the twenty-year period.

Table 2.8 Neighborhood Mobility through Ecological Categories

A. Neighborhood Mobility, 1970–80			1980 Category		
1970 Category	Middle-Class	Yuppie	Working-Class	Underclass	Total
Middle-Class	270	8	50	14	**342**
Yuppie	9	63	13	2	**87**
Working-Class	52	19	152	151	**374**
Underclass	0	0	2	20	**22**
Total	**331**	**90**	**217**	**187**	**825**

B. Neighborhood Mobility, 1980–90			1990 Category		
1980 Category	Middle-Class	Yuppie	Working-Class	Underclass	Total
Middle-Class	248	41	31	11	**287**
Yuppie	0	90	0	0	**90**
Working-class	26	40	82	69	**217**
Underclass	4	4	3	176	**187**
Total	**278**	**175**	**116**	**256**	**825**

C. Neighborhood Mobility, 1970–90			1990 Category		
1970 Category	Middle-Class	Yuppie	Working-Class	Underclass	Total
Middle-Class	211	44	*50*	37	**342**
Yuppie	2	77	4	4	**87**
Working-Class	*65*	*54*	*62*	*193*	**374**
Underclass	0	0	0	22	**22**
Total	**278**	**175**	**116**	**256**	**825**

gentrifying yuppie and ghetto underclass neighborhood types—gained increasing shares of the city's neighborhoods over time. As first identified by W. J. Wilson (1987), most of the growth in the ghetto underclass category occurred during the 1970s, when it expanded from 22 tracts (3 percent of the city total) in 1970 to 187 tracts (23 percent of the city total) in 1980. The gentrifying yuppie category also expanded over time, particularly during the 1980s, when its share grew from 11 to 21 percent of the city's tracts (i.e., from 90 to 175 tracts). Thus, the ecological structure of Chicago's neighborhoods became more polarized over time, with the erosion of the middle ground occupied by transitional working-class areas and the concomitant emergence of more impoverished areas in the 1970s and gentrifying neighborhoods during the 1980s. That the 1970s were more crucial in the formation of ghetto underclass areas reflects the profound industrial restructuring that occurred in response to the oil embargo "price shocks," which by the 1980s had precipitated multifaceted adjustment strategies. That is, the 1970s represented a period of major structural changes in economic relations that reverberated on all facets of social life, while economic changes during the 1980s accentuated inequalities by awarding rising returns to skill and by reinforcing the trends set in motion during the 1970s.

Another important insight is the tendency for neighborhoods to remain in the same ecological category over time (i.e., to fall on the main diagonal of the mobility tables). This pattern is particularly pronounced among the gentrifying yuppie and ghetto underclass neighborhoods, and it is evident to a lesser extent in middle-class areas. Transitional working-class neighborhoods were the most fluid of all clusters. Panel C of table 2.8 shows that only 17 percent (62 out of 374) of the neighborhoods in this category in 1970 remained there in 1990. Thus, these neighborhoods served as turning points in neighborhood trajectories. Among the working class there was much "transition." Many neighborhoods downgraded and joined the ghetto underclass category, and a nontrivial share upgraded and joined either the stable middle-class or the gentrifying yuppie category. Most of the downgrading of transitional working-class neighborhoods occurred during the 1970s, when 40 percent (151 out of 374) became ghetto underclass areas, while most of the upgrading took place over the following decade, when 33 percent (66 out of 217) of the transitional neighborhoods moved either to the stable middle-class or gentrifying yuppie category. These transitions were somewhat counterbalanced by a smaller inflow of neighborhoods into working-class areas, particularly from the stable middle-class neighborhood type. In fact, 43 percent (50 out of 116) of

Table 2.9 Descriptive Statistics by Pattern of Neighborhood Change

Pattern of Neighborhood Change	1970		1980		1990	
	Mean	(Std Dev)	Mean	(Std Dev)	Mean	(Std Dev)
Middle-class to working-class						
%Black	0.3	(1.0)	7.0	(16.3)	21.7	(29.9)
%Non-Hispanic white	93.7	(6.1)	56.9	(21.4)	27.4	(18.7)
%Hispanic	6.2	(7.2)	31.4	(18.6)	45.2	(24.2)
SES Index	0.37	(0.16)	0.02	0.22	−0.03	(0.33)
Resid stability index	0.18	(0.40)	−0.08	0.54	−0.16	(0.39)
Stable working-class						
%Black	5.5	(16.8)	5.1	(14.1)	6.6	(14.0)
%Non-Hispanic white	60.5	(22.0)	30.8	(18.4)	20.5	(15.1)
%Hispanic	34.5	(19.3)	62.0	(21.0)	70.5	(20.5)
SES Index	0.02	(0.20)	−0.40	(0.28)	−0.23	(0.30)
Resid stability index	−0.50	(0.35)	−0.55	(0.42)	−0.35	(0.36)
Working-class to middle-class						
%Black	48.8	(39.8)	59.9	(44.4)	61.8	(45.0)
%Non-Hispanic white	40.0	(33.0)	20.9	(28.5)	16.2	(25.0)
%Hispanic	11.9	(18.5)	18.5	(28.0)	19.6	(31.0)
SES Index	−0.27	(0.39)	−0.21	(0.42)	0.03	(0.39)
Resid stability index	−0.27	(0.54)	0.54	(0.84)	0.66	(0.69)
Working-class to yuppie						
%Black	18.3	(30.1)	20.1	(28.2)	18.9	(25.9)
%Non-Hispanic white	58.4	(25.9)	47.4	(21.9)	54.5	(24.8)
%Hispanic	20.5	(13.4)	24.7	(17.0)	17.5	(15.0)
SES Index	0.14	(0.27)	0.15	(0.47)	0.67	(0.54)
Resid stability index	−0.80	(0.49)	−0.73	(0.49)	−0.79	(0.46)
Working-class to underclass						
%Black	78.3	(35.6)	84.9	(29.3)	86.8	(27.2)
%Non-Hispanic white	13.8	(24.3)	5.2	(11.4)	2.9	(7.3)
%Hispanic	7.2	(15.4)	9.4	(20.8)	9.7	(22.5)
SES Index	−0.25	(0.32)	−0.99	(0.40)	−1.04	(0.42)
Resid stability index	−0.54	(0.51)	−0.13	(0.53)	−0.09	(0.58)

Source: U.S. Census 1970–90, City of Chicago
Note: Italicized cells refer to Panel C of Table 2.8.

the neighborhoods that were working-class areas in 1990 began the 1970s in the stable middle-class category.

To understand the implications of neighborhood change for different demographic groups, we inspected in closer detail the italicized cells in Panel C of table 2.8. Descriptive statistics for these cells, which are identified by their coordinates in table 2.8, are shown in table 2.9. We report racial and ethnic composition, as well as an index of socioeconomic status and an index of residential stability. These indices provide summary indicators of most variables listed in table 2.7 that were used in the cluster analysis. Positive values reflect higher levels of socioeconomic status and residential stability, respectively, and the scale is measured in standard deviation units.

The vast majority of neighborhoods that were in the transitional working-class category in 1990 followed one of two modal paths: either they moved there from the stable middle-class category (cell 13), or they began the period (1970) in the transitional working-class category and remained there in 1990 (cell 33). Both trajectories are important for understanding the emergence of working-class Hispanic neighborhoods in Chicago and for understanding how color, that is, group membership, circumscribes opportunity in Chicago. The top panel of table 2.9 reveals that transitions from stable middle-class to transitional working-class neighborhoods were associated with a rapid growth of the Hispanic population. In these areas, the Hispanic share rose from 6 percent in 1970 to 45 percent in 1990. At the same time, the non-Hispanic white composition of these neighborhoods declined dramatically, from 94 percent in 1970 to 27 percent in 1990. This means that many Hispanic neighborhoods evolved out of formerly white, stable, middle-class areas of Chicago. This transition was accompanied by moderate declines in both socioeconomic status and residential stability. The second modal path associated with Hispanic neighborhood change is represented by neighborhoods that remained in the transitional working-class category throughout the period under consideration. These neighborhoods, which increased from 34 percent Hispanic in 1970 to 70 percent in 1990, were early ecological footholds for Chicago's Hispanic population. Neighborhoods that followed this trajectory were of lower socioeconomic status and experienced high levels of residential turnover compared to those emerging from the stable middle-class category (compare the first and second panels of table 2.9).

Although most neighborhoods that evolved from transitional working-class areas downgraded to underclass areas by 1990, a substantial number also upgraded by moving either to the stable middle-class or gentrifying yuppie category. Given the tendency to equate the ghetto underclass with black neighborhoods, it is pertinent to ask whether all black neighborhoods that were on the cusp of ghetto poverty in 1970 did eventually tip and become "underclass" areas, or whether there was also a substantial countertrend toward neighborhood upgrading that was ignored in the urban poverty literature. Our results show that upgrading did occur among both black and white neighborhoods. In terms of race-specific patterns, black neighborhoods that upgraded were more likely to join the stable middle-class category (cell 31), while white neighborhoods were more likely to enter the gentrifying yuppie category (cell 32).

The third and fourth panels of table 2.9 shed more light on these paths of neighborhood change. Neighborhoods that moved out of the transitional working-class category and into the stable middle-class category were nearly 50 percent black in 1970, and that percentage rose to more than 60 percent by 1990. Likewise, neighborhoods in this cell witnessed overall declines in white population, from 40 percent in 1970 to 16 percent in 1990. These neighborhoods lost socioeconomic ground over time, but they remained slightly above the citywide averages in 1990. Neighborhoods that moved from working-class to middle-class composition also became more residentially stable over time. White neighborhoods dominated the second mode of upgrading, which was characterized by neighborhoods that moved from transitional working-class to gentrifying yuppie areas. The socioeconomic status of these neighborhoods increased substantially during the 1980s, and residential stability remained at very low levels throughout the entire period.

Table 2.9 basically confirms a common finding in studies of urban poverty, namely, that the process of ghetto formation was almost exclusively confined to blacks. Neighborhoods that moved from the transitional working-class category to the underclass category became increasingly homogenous in racial composition over time. These neighborhoods witnessed dramatic declines in socioeconomic status, particularly during the 1970s. Our analysis of Chicago neighborhoods shows that the dramatic spread of ghetto poverty, particularly during the 1970s, was also accompanied by several additional ecological trends that transformed the social landscape of the city. One of these is the spatial concentration of affluence, exemplified by the spread of gentrifying yuppie neighborhoods during the 1980s.

The net effect of these two transformations was to heighten the spatial polarization of inequality in Chicago neighborhoods. Another trend was the erosion of the transitional working-class areas that occupied the middle ground in Chicago's ecological structure. However, the city's burgeoning Hispanic population appears to have transformed these areas into a new ecological niche that shows signs of economic vitality as well as endurance. Finally, a more subtle temporal shift was the gradual decline of stable middle-class areas, along with their ecological transformation into black middle-class enclaves. Mapped against high levels of residential segregation, and to the extent that social and economic opportunities are spatially delimited, these changes portend highly unequal life chances for Chicago's black, white, Mexican, and Puerto Rican families. Specifically, they suggest high joblessness and welfare participation among blacks, who became increasingly confined to the most disadvantaged areas, and high levels of economic activity among Mexicans, whose neighborhoods expanded even while they retained a solid working-class character. In fact, the evidence we present in the following chapters confirms these expectations, but our approach emphasizes the vulnerability of groups that experienced material disadvantage during childhood and adolescence.

Concluding Thoughts

At the turn of the twenty-first century, Chicago is a city in transition, experiencing population losses and marked changes in its racial and ethnic composition. Three demographic trends—the exodus of whites from the inner city, the rapid growth of the Latino populations, and a slowdown in the growth of the black population—have conspicuously altered the city's racial and ethnic landscape. These changes were accompanied by equally dramatic changes in the size of the employment base and the industrial composition of employment, and within the city limits by spatial polarization of inequality across neighborhoods. Specifically, Chicago lost more than two hundred and eighty thousand manufacturing jobs between 1970 and 1990, and those losses were only partially offset by the addition of a hundred and sixty thousand nonmanufacturing jobs, mostly in low-wage service employment. The coupling of these changes with a decentralization of employment, away from the central city toward the suburban city fringe and the adjacent counties comprising the metropolitan statistical area, delimited economic opportunity in decisively unequal ways for black, white, and Latino populations. In addition, against the backdrop of high levels of racial and ethnic residential segregation, the confluence of several factors, notably the spatial redistribution of job opportunities and a drop in demand for unskilled labor, accentuated the deleterious impact of industrial restructuring for Chicago's black and Latino populations.

However, our overview of social and economic trends in Chicago exposed several anomalies that warrant further examination. Specifically, the descriptive profiles of racial and ethnic differences in educational attainment identified Mexicans as the most disadvantaged group, yet blacks and Puerto Ricans fared worse economically, as indicated by their lower rates of labor force participation and higher rates of unemployment, poverty, and welfare participation. We also uncovered several paths of neighborhood change that altered the social ecology of Chicago, increasing in particular the spatial polarization of the city's neighborhoods. These ecological changes are socially significant not only because they dramatically altered the city's social ecology, but also because they did not affect all demographic groups uniformly. As an important corrective to the urban poverty literature, we noted that among black neighborhoods, the process of neighborhood change is not unidirectional. However, when neighborhoods do upgrade, they appear to follow race-specific patterns. Whereas upgrading white neighborhoods tend to follow the path of gentrifying yuppie neighborhoods, emerging black middle-class neighborhoods take on an ecological character more consistent with stable middle-class areas of the city.

Our final observation concerns the understudied process of Hispanic neighborhood change. Specifically, we showed that Hispanic populations followed the classic ecological pattern of establishing an early foothold in transitional working-class areas of Chicago, but many Hispanic neighborhoods succeeded white populations in formerly stable middle-class areas of the city. The trajectories of neighborhood change in areas populated by Hispanics are socially significant because, unlike the poor, black ghettos, they hold promise of social mobility across, if not within, generations. These differences in the ecological experiences of blacks and Hispanics, especially Mexicans, yield insights about a puzzle that underlies our concerns about the relative influence of achievement and ascription in shaping socioeconomic inequality in Chicago. Why have neighborhoods populated by recent Mexican immigrants remained viable, working-class areas, while black neighborhoods were more likely to become underclass areas in the face of massive industrial restructuring? We address this question in subsequent chapters, following a description of the project and data sources used for the statistical analyses.

CHAPTER THREE

The Study Population

This chapter provides a general overview of the Urban Poverty and Family Life Study and its component surveys, which we analyze extensively in the ensuing chapters. The Urban Poverty and Family Life Study is a multiyear, interdisciplinary research project conducted at the University of Chicago between 1985 and 1990.[1] As the centerpiece of the study, the Urban Poverty and Family Life Survey (UPFLS) was designed to describe and understand the lives of black, white, Mexican, and Puerto Rican families living in impoverished inner-city neighborhoods. The Social Opportunity Survey (SOS) complemented the main survey with in-depth, semistructured interviews conducted with a subset of respondents to the general survey. In addition, the complete study included a survey of 187 employers located in or near inner-city neighborhoods and extensive ethnographic research consisting of community ethnographies and individual biographies.

In this monograph we draw on both the UPFLS and the SOS to examine race and ethnic differences in family, welfare, and work experiences of inner-city residents. Throughout the book we also selectively compare Chicago's inner-city residents with representative samples of urban black, white, Mexican, and Puerto Rican families using the National Survey of Families and Households (NSFH). Comparisons of parents residing in extremely poor and moderately poor urban neighborhoods with urban

1. William J. Wilson was the principal investigator. The history of this complex project has been detailed by Krogh (1991).

parents nationally permit us to address various facets of the core question driving the underclass debate, namely, whether and in what ways the behavior of the inner-city poor differs from that of the urban poor in general.

The chapter describes the three surveys analyzed herein and characterizes the study population. Following a description of the sampling framework and survey content of Chicago surveys, we address several points of controversy about the study and clear up some misunderstandings about the UPFLS. Subsequently, we describe the National Survey of Families and Households, highlighting its comparability with the UPFLS and our strategy to identify the urban poor using the national probability survey. Finally, we systematically compare the UPFLS, SOS, and NSFH by providing statistical portraits of the three samples.

The Chicago Urban Poverty Surveys

This section describes the design and content of the Urban Poverty and Family Life Survey (UPFLS) as well as its companion, the Social Opportunity Survey (SOS).[2] In addition to discussing the strengths and limitations of these surveys for investigating the family, welfare, and employment behavior of parents residing in poor, inner-city neighborhoods, we consider issues of generalizability to other urban populations and elaborate a strategy to broaden statistical inference to the urban poor in general.

Sampling Framework

The UPFLS is based on a multistage, stratified probability sample of parents aged 18–44 who resided in poverty census tracts. Completed interviews, which averaged just under two hours, were obtained from 2,490 respondents, including 1,183 blacks, 364 whites, 489 Mexicans, and 454 Puerto Ricans.[3] The overall response rate of 79 percent ranged from 74 percent (whites) to 83 percent (black parents). Response rates for

2. The National Opinion Research Center was commissioned both to conduct the Chicago survey (draw the sample, train and supervise interviewers, and prepare a machine-readable data file) and to work with the project team in the design of the survey instrument. The tasks of drawing representative samples of poor, inner-city residents, achieving acceptable response rates, and ensuring high-quality data were challenges that only the best survey research organizations could manage.

3. Interviews were conducted in English and Spanish, and the latter were administered by bilingual interviewers.

Mexicans and Puerto Ricans were 78 and 77 percent, respectively.[4] Although the original survey was conceived as a study of parents, 156 blacks who were not parents were also interviewed. In the interest of enhancing comparability across ethnic groups, we restrict the empirical analyses to parents, which yields a final analysis sample of 2,334, including 811 men and 1,523 women.

The study design called for respondents living in poor neighborhoods, liberally defined as census tracts where at least 20 percent of families had incomes below the federal poverty line. When the survey was designed, the census data (from 1980) showed that 35 percent of Chicago's population resided in such census tracts, including about 62 percent of all blacks, 47 percent of Mexicans, 54 percent of Puerto Ricans, and 8 percent of non-Hispanic whites. Because not all racial and ethnic groups were adequately represented in these neighborhoods, obtaining the targeted sample sizes for each group (1,200 blacks and 500 each of Mexican, Puerto Rican, and non-Hispanic whites) required a stratified sample. Details about the sampling procedure are described in appendix A.

Content of the Urban Poverty and Family Life Survey

The Urban Poverty and Family Life Survey is comprehensive in its coverage of the social and family characteristics of respondents. In addition to standard questions on respondents' family background, education, and military experience, the survey includes detailed information about household composition. Extensive life history calendars recorded birth, work, welfare, and personal relationship histories on a retrospective basis. This data collection strategy minimizes left censoring, that is, unobserved events that occurred before the survey began, but increases reporting error through recall of events that occurred in the distant past. Because retrospective income data are notoriously unreliable, earnings data were obtained only for current job rather than for the entire work history. Respondents were queried about current monthly household income using a precoded categorical scale and also about various household income sources (i.e., labor income, public assistance, food stamps, social security or veterans' benefits, and unemployment compensation). The survey solicited information about neighborhood and housing characteristics,

4. Because no formal analysis of nonresponse rates was conducted, the types of biases introduced are unknown. Field operations indicated that nonresponse arose largely because of difficulties locating respondents and especially because of the high mobility of some inner-city populations.

respondents' health status, social networks, major life events, and attitudes about work, welfare, economic opportunity, and family life. Also available are several open-ended questions about means of support and attitudes toward work, welfare, and family life, from which we draw selectively to interpret our results.

Most of the statistical analyses for chapters 4, 5, and 6 are based on birth and marriage, welfare, and employment histories. Chapter 4 exploits family background information along with the birth and marriage histories to characterize the early life-course experiences of survey respondents. In addition to considering two aspects of the adolescent life course, namely, premature school withdrawal and adolescent fertility, we examine family formation patterns by combining the marriage and fertility histories to portray alternative pathways to family life. Welfare histories, the focus of chapter 5, contain information for up to four spells of public assistance for each respondent, specified by type of grant (i.e., AFDC, GA, SSI, or other). In addition, respondents indicated whether their family of orientation ever received welfare during their first fourteen years of life. Chapter 6 analyzes the rich labor force histories and current information on reservation wages to address questions about racial and ethnic differences in willingness to work as well as more standard questions about the determinants of labor supply.

The Social Opportunity Survey

The UPFLS was supplemented by a "mini-survey," the Social Opportunity Survey (SOS), an in-depth, open-ended interview conducted with a small subset ($N = 161$) of respondents from the UPFLS. Respondents for the SOS were selected from the pool of individuals who responded to the main survey using quota sampling. Completed surveys were sorted, and potential respondents were selected from the poorest census tracts in the proportion of 4:2:2:1, respectively, for black, white, Mexican, and Puerto Rican respondents. Within census tracts and demographic groups, equal numbers of men and women were selected from among the employed and unemployed, and from among the married and unmarried. Because of difficulties locating potential sampled respondents, only half the anticipated number of SOS interviews were completed.

The SOS was not designed as a random subsample of the UPFLS. Nevertheless, SOS respondents span the racial, ethnic, and gender variation of the general survey. Responses to the mini-survey provide richly textured information that aids in interpreting the statistical results from the general

Figure 3.1 Questions Analyzed from the Social Opportunity Survey

Section 1: Opportunity and Mobility

- Is the United States a land of opportunity where people can advance and where they get what they deserve?
- Do people have more opportunities to get ahead nowadays or in the past?
- What is the main way to get ahead in Chicago today?

Section 2: Education and Expectations of Children

- Some people think that education is the best way to get ahead. What is your opinion?
- Is your life better than your parents' life?
- Has family life in Chicago changed much in the last twenty years?

Section 3: Work Experience and Opinions

- What do you consider to be a "good job"? What is most important to get a good job in Chicago? Do you think you will get a good job like that?
- Who are the people who have the good work in this city?
- Are there enough jobs for all the people in this country? In Chicago? If not, why not? What should be done to improve the situation?
- If someone is willing to work for $3.35 an hour, can he get a job right now?[a]

Section 6: Social Classes

- What does the phrase "the American Dream" mean to you?
- There is much talk these days about an "underclass" in the United States. Does this word mean anything to you?

[a] $3.35 was the minimum wage when the study was done.

survey. All SOS respondents can be matched to the UPFLS responses to contextualize responses in social, demographic, and economic histories. That is, the verbatim quotes can be attributed to individuals with different family, welfare, and work histories.

The SOS used open-ended questions to solicit more in-depth responses about several topics covered in the main survey, namely, opportunity and mobility, educational expectations for children and family life, work experience and opinions about job opportunities, and the American Dream. Figure 3.1 summarizes the items we analyzed from among the full set of questions included in the mini-survey. Our selection was guided by the theoretical and substantive concerns of the family, welfare, and employment chapters.

Interviews for the mini-survey lasted about one hour; they were tape-recorded and transcribed verbatim.[5] Our interest in family, welfare, and

5. We used the ASKSAM data manager to access the verbatim responses, but resorted to more conventional methods—that is, detailed reading and rereading, coding and recoding—for interpretation of the quantitative analyses.

work directed our attention to a subset of questions about opportunity and mobility, family life, work experiences, and meanings of the underclass and the American Dream.

Strengths and Limitations of UPFLS

From its inception, the Urban Poverty and Family Life Study was designed to focus on poor, inner-city populations, but concerns about the survey's limited generalizability to nonpoor residents and to nonparents remained a constant source of tension throughout the project. Since the project began, the merits of supplementing the proposed sample of parents residing in poor places with a subsample of middle-class neighborhoods or extending the sampling frame to the entire city of Chicago have been debated. However, cost and time considerations dictated that the initial project purposes be honored because sampling in poor neighborhoods was the only cost-efficient strategy to study social behavior in low-income environments while ensuring a sufficient number of observations to permit intensive analysis of specific racial and ethnic groups. Unfortunately, this feature of the study has been greatly misunderstood.

Poor Places versus Poor People

The main drawback to fielding a study that concentrates primarily on poor neighborhoods is that it begs the question of whether and in what ways the behavior of parents who do not reside in poor neighborhoods differs from the behavior of those who do. However, not all poor people reside in poor neighborhoods; in fact, the majority of the poor do not live in concentrated poverty areas. Similarly, not all people residing in poor places are themselves poor, although many are, and in the poorest neighborhoods most are. This distinction bears crucially on our ability to draw inferences from the UPFLS that are generalizable to a broader population of poor people.[6] Thus, while we recognize the limitations of a sampling framework that focuses on poor neighborhoods, it bears emphasis that the UPFLS is not a sample of poor people. This feature would impose much

6. Inclusion of residents from tracts with very low poverty in the survey would permit generalization to the entire city if appropriate weights are applied. Although it would be possible to generalize to the entire city of Chicago using a modified weighting scheme, this is not our purpose, nor was it the goal of the study.

more severe limits on the possibilities for statistical inference.[7] In our opinion, much of the controversy that has surrounded this survey stems from the tendency of critics to conflate the distinction between poor places and poor people (see Tienda 1991).

Although poor people obviously have a greater probability of selection in the highest-poverty census tracts, respondents were randomly selected within tracts, and hence the sample contains considerable socioeconomic variability. The high income and racial segregation of Chicago resulted in the inclusion of highly affluent and desperately poor respondents who reside in neighborhoods where at least 20 percent of all residents were poor. It is also noteworthy that although residents of nonpoor neighborhoods were underrepresented relative to those in ghetto-poverty neighborhoods, they were not completely excluded from the sample (for details, see appendix A). In other words, some individuals included in the sample did reside in nonpoor neighborhoods (i.e., less than 20 percent poor residents) at the time of the survey. The inclusion of respondents from nonpoor neighborhoods in the sample was the result of both (1) a supplemental oversample of whites and Hispanics, which had to draw upon residents of less poor neighborhoods in order to achieve adequate representations of these groups; and (2) residential mobility taking place among some respondents who were screened for the survey while living in a poor neighborhood but subsequently moved to a less poor neighborhood before the time of the actual survey.

While the survey was in the field, the definition of "underclass" areas evolved from those where 20 percent of residents were poor to neighborhoods where at least 40 percent of residents had incomes below the poverty line (Ricketts and Sawhill 1988; Ruggles 1990). This means that the neighborhood variation included in the UPFLS is much broader than the eventual focus of many "underclass" studies. Using the typology presented in chapter 2, the UPFLS represents primarily ghetto underclass and transitional working-class neighborhoods, and, to a lesser extent, gentrifying yuppie neighborhoods. Stable middle-class neighborhoods are essentially excluded, but they are less relevant for the study of urban poverty.

7. Because observations were randomly selected within sampling strata, using maximum likelihood methods, as we do, produces unbiased and efficient estimates. This is critical for reliable statistical inference (Maddala 1983, 165–74).

Parent Focus

Another issue warranting further discussion is the decision to survey parents, to the exclusion of nonparents. While this was theoretically justified in terms of the overall aims of the Urban Poverty and Family Life Study, some critics argued that this decision could introduce selection biases in the study of some behaviors, such as work, educational achievement, and marital histories. This is because the act of becoming a parent itself may alter these activities in some systematic way that cannot be observed because of the exclusion of a nonparent control group.

To such critics we offer several responses. First, the centrality of family structure issues in the genesis of the Urban Poverty and Family Life project justifies the focus on parents both theoretically and substantively. It makes little sense for a study about family life to study nonparents if the purpose is to understand the rise in mother-only families, for example. Second, the inclusion of detailed, retrospective birth and marriage histories obviates this problem to a considerable degree. This is because the retrospective birth and marriage histories permit us to portray entry to family life via birth or marriage and to evaluate the consequences of these alternative pathways to parental status.

Some critics argued that using parenthood as a major sampling criterion means that any life event after the conception of the first child can be treated as a dependent (or outcome) variable in statistical analysis, but that any event up to and including that event could only be treated as an independent variable. In other words, analyses of welfare participation should be unbiased because AFDC receipt is conditional on having a child, but employment behavior may differ before and after a birth. Clearly, selection on parental status is less consequential for analyses of AFDC recipiency, where parental status is a condition for receipt. For analyses of employment, we ensure that time-varying regressors are dated prior to the outcome of interest.

Generalizations from Chicago to Other Urban Populations

Thus far we have discussed two features of the UPFLS sampling framework that constrain the bounds of its generalizability: its focus on parents and its overrepresentation of residents living in low-income neighborhoods. On a more basic level, the ability to generalize from this sample is also limited by confinement to a single location. On many social and economic indicators, such as residential segregation, poverty concentration, and economic structure, Chicago may exhibit extreme values relative to other major industrial cities. Thus, the findings based on this single-site survey cannot be generalized to residents of poor neighborhoods in other cities. Yet this limitation also is a strength because many ideas about how the social and economic consequences of concentrated urban poverty differ from more generalized experiences with poverty were derived from firsthand accounts of Chicago (e.g., W. J. Wilson 1987). Therefore, Chicago is a fitting test case for evaluating the merit of these ideas.

In order to determine how similar the UPFLS sample is to other urban populations, we compare Chicago's inner-city parents with a national urban sample of parents from the National Survey of Families and Households (NSFH) and also with a subsample of low-income urban parents. This comparison illustrates whether and how the Chicago residents differ from urban residents generally and from the poor in particular. We next turn to a description of the national survey.

The National Survey of Families and Households

The National Survey of Families and Households presents a particularly convenient benchmark against which to compare the Chicago Urban Poverty and Family Life Survey. Both surveys were fielded at approximately the same time and contain many identical questions (both principal investigators shared instruments), which greatly enhances cross-survey comparisons. Both the UPFLS and the NSFH were based on cross-sectional designs, which include retrospective sequences on early adult experiences. The retrospective event histories allow us to portray how early life-course experiences eventuate in contemporary adult outcomes. Both surveys contain detailed information about respondents' family backgrounds and current income status, and both surveys obtained retrospective information about whether respondents had ever been married, had ever worked, or had ever experienced a birth during adolescence (i.e., before age nineteen). In addition, both instruments solicited information about whether respondents were reared by two parents and whether their parents received public assistance when respondents were growing up. Because the NSFH is a more general-purpose survey about families and households, and not a study of poverty and welfare dynamics, information about employment and welfare histories is more limited in the national sample. These differences in the purposes of the two surveys inevitably limit the detail of some comparisons, but not our broad analytical objectives and especially not the ability to consider whether and in what ways the behavior of the inner-city poor differs from that of the poor in general.

Our comparisons provide national urban benchmarks to anchor contrasts of racial and ethnic groups in ways that both inform the bounds of generalizability and clarify whether and in what ways black, white, Mexican, and Puerto Rican inner-city parents behave differently from their urban counterparts nationally. Because this survey has been in the public domain since it was collected and is generally well known (see Sweet and Bumpass 1987), we describe its design in less detail than the UPFLS, emphasizing our construction of subsamples that enhance cross-survey comparisons.

The NSFH, while it is a national probability sample of 13,017 respondents, consists of two main parts. The main sample includes 9,643 respondents who represent the noninstitutional U.S. population aged nineteen and older. In addition, minority groups, cohabiting adults, and single parents were oversampled. These features also suit our analytical goals because they ensure adequate samples of minority women. For each household, one person was randomly selected as the main respondent. The spouse, partner, or other nonspouse adult responded to a shorter, self-administered questionnaire.

Urban Subsamples of the NSFH

To increase comparability with the UPFLS sample, which is based on parents residing in inner-city neighborhoods, we restricted our analysis of the NSFH sample to parents aged 19 to 45 who resided in a central city. The resulting sample includes 1,887 women (529 blacks; 1,095 whites; 201 Mexicans; and 62 Puerto Ricans) and 947 men (213 blacks; 613 whites; 101 Mexicans; and 20 Puerto Ricans). To further sharpen comparisons between poor people and residents of poor places, for some analyses we used a subsample of low-income, central-city respondents. The low-income stratum consists of parents with household incomes equal to or below 1.5 times the poverty line.[8] This stratification enables us to compare

8. The NSFH does not provide poverty status or household income data for women who are not heads of household, even if they have children. This design feature affected 3.3 percent of mothers in our urban subsample. For these women, we estimated poverty status using information about their household structure, their own and their spouses income, and their welfare participation. This information was used only in sample stratification, that is, assignment of low-income status. We conducted diagnostic tests to evaluate the sensitivity of our results to the allocation decisions and determined that results were not altered by the allocation decisions.

the Chicago inner-city sample against a national urban population as well as a low-income urban population, which is crucial for addressing whether and how the inner-city poor differ from the poor in general.[9]

According to 1990 census figures, Chicago's population was approximately 40 percent black, 20 percent Hispanic (overwhelmingly Mexican and Puerto Rican), and about 40 percent white and other races, including Asians. The inner-city sample overrepresents blacks and underrepresents whites relative to the citywide averages, but this is because blacks are far more likely than whites to live in poor neighborhoods. However, the Hispanic share of the UPFLS is relatively similar to the group share of the citywide population.

Table 3.1 reports unweighted sample counts for both surveys, including the poor substratum of the national sample. Also reported are sample differences in the racial, ethnic, and sexual composition of the three samples compared. Obviously, the high minority concentration in Chicago combined with the deliberate scheme to oversample blacks and Hispanics renders the UPFLS a largely minority sample (93 percent). Nationally, nearly half of urban poor residents are white, 26 percent are black, and an additional 26 percent are Hispanic. Among all urban central-city residents, 70 percent are white and 30 percent black.

Comparability with the UPFLS

Although very similar in their content, the two surveys are notably different in some ways that constrain the types of comparisons we can make. For example, the UPFLS recorded welfare episodes, marriages, employment, and fertility by year and month on a life-history calendar. This enables us to determine with reasonable precision the beginning and ending dates of spells, the number of spells, and whether any were in progress at the time of the interview. For welfare episodes, the NSFH provides information only about *recent* welfare experiences, that is, episodes of AFDC receipt that occurred during the five years preceding the survey. For each of the five years preceding the survey, NSFH respondents indicate whether they received welfare in each year, but they do not report the beginning or ending dates for each episode. Because the number and

9. In this regard, it is noteworthy that the low-income stratum from the NSFH is a truncated sample because all households with income higher than 1.5 times the poverty line were excluded. In cases where the dependent variable is household income, the OLS regression estimates are biased, but estimates of welfare participation based on maximum likelihood methods are unbiased (see Maddala 1983, 165–74 for a detailed discussion of the problem).

Table 3.1 Distribution of UPFLS and NSFH Respondents by Ethnicity and Sex (Weighted Percents, Unweighted Counts)

	Chicago UPFLS		U.S. Urban NSFH		U.S. Urban Poor NSFH	
	%	*N*	%	*N*	%	*N*
Blacks						
Men	22.4	308	6.1	213	5.5	39
Women	45.7	719	11.5	529	21.0	218
Total	68.1	1027	17.6	742	26.5	257
Whites						
Men	2.3	127	30.5	613	15.3	57
Women	4.5	237	39.6	1095	31.7	243
Total	6.8	364	70.1	1108	47.0	300
Mexicans						
Men	8.2	228	4.3	101	6.9	34
Women	9.3	261	5.9	201	15.0	107
Total	17.5	489	10.2	302	22.0	141
Puerto Ricans						
Men	2.4	148	0.7	20	1.2	7
Women	5.1	306	1.3	62	3.3	32
Total	7.5	454	2.0	82	4.5	39
Grand Total[a]	100	2334	100	2834	100	737

Source: UPFLS and NSFH, Urban Percent Subsample

[a] May not sum to 100 due to rounding.

duration of spells were not ascertained in the NSFH, we cannot address questions about the chronicity of welfare participation for the national survey nor its low-income substratum. However, our intention in drawing cross-survey comparisons is not to compare every aspect of family structure, welfare participation, and employment behavior, but to situate Chicago's residents against alternative benchmarks so that the behavioral uniqueness of the inner-city poor, if it exists at all, is clear.

Another important difference between the surveys is the availability of information about neighborhoods in the UPFLS that is unavailable in the NSFH. The need for this information was driven by hypotheses about how social environments influence behavior and, more specifically, how residence in high-poverty areas was conducive to high rates of non-normative behavior, notably, teenage childbearing, premature school withdrawal, delinquency, prolonged joblessness, and chronic welfare dependence. For the NSFH, geo-codes are restricted to the county level and therefore cannot provide the nuanced detail about neighborhood environments available in Chicago.[10]

In the UPFLS, neighborhood characteristics correspond to the respondents' residence at the time of the survey. However, the UPFLS lacks time variation in neighborhood characteristics. Thus, despite the availability of rich data about neighborhood environments in the UPFLS, it is unfortunately impossible to relate these circumstances to the life histories because the survey lacks information about the respondents' exposure to the neighborhoods where they were surveyed. This is not a problem for respondents who lived their entire lives in the same neighborhood where they were interviewed, but only a small minority of the sample met this condition. Therefore, it is not possible to relate neighborhood characteristics to longitudinal analyses of family, work, and welfare experiences of inner-city residents. Nevertheless, the fact that the NSFH lacks rich neighborhood characteristics does not greatly compromise the comparative analyses of changes in the family life, work, and welfare of inner-city parents.

Empirical Profiles of the UPFLS and NSFH Surveys

This section describes and compares the respondents of both the UPFLS and NSFH in terms of key social and demographic characteristics. Our descriptions are necessarily selective and designed to frame the themes investigated in the chapters that follow. They also serve to identify average differences (and similarities) among urban parents nationally, low-income urban parents nationally, and Chicago's inner-city parents. Here we introduce the monograph's core theme of racial and ethnic differences by drawing comparisons among black, white, Mexican, and Puerto Rican men and women.

Table 3.2 summarizes characteristics of the UPFLS sample by ethnicity and sex. Given the age-selection criteria, there were minor age differences among the groups, although whites were two to three years older than their minority sex counterparts. However, despite the sample restriction to parents, there were large group differences in marital status. Nearly half of black women had never married compared to one in three Puerto Rican, one in five white, and 7 percent of Mexican-origin women. The reported shares of never-married fathers were much lower, although it is conceivable that these rates are understated (Stier and Tienda 1993).

10. For example, in the NSFH, there is one geo-code for Cook County, but no subarea codes to distinguish poverty from nonpoverty neighborhoods within the City of Chicago.

Table 3.2 Sample Characteristics by Ethnicity and Sex: Urban Poverty and Family Life Survey (Means and Percents)

	Blacks		Whites		Mexicans		Puerto Ricans	
	Men	Women	Men	Women	Men	Women	Men	Women
Age (years)	31.8	31.4	35.5	33.7	33.5	31.4	33.2	31.7
(s.d.)	(7.3)	(7.1)	(6.1)	(3.9)	(5.9)	(6.8)	(6.7)	(6.8)
Education								
H.S. graduate (%)	54.8	58.0	67.1	63.6	38.3	35.6	31.0	38.7
Education (years)	12.0	11.9	13.1	12.1	7.0	7.2	9.7	9.9
(s.d.)	(2.1)	(1.9)	(2.9)	(2.6)	(3.6)	(3.8)	(3.0)	(2.8)
Marital Status (%)								
Currently married	41.5	23.2	74.7	53.8	85.3	68.7	66.6	36.3
Currently divorced/sep	16.4	30.3	20.5	26.6	7.6	16.2	9.3	31.4
Never married	42.1	46.5	4.8	19.6	7.1	15.1	24.1	32.3
Current Employment Status (%)								
Employed	69.7	44.1	87.0	52.6	93.1	50.9	76.8	34.1
Unemployed	12.9	6.4	4.3	3.3	1.7	1.9	7.6	1.3
Out of labor force	17.4	49.5	8.7	44.1	5.2	47.2	15.6	64.6
Welfare Participation (%)								
Currently on aid	22.3	52.0	6.2	29.7	2.6	13.6	11.6	53.7
Ever received aid	46.1	81.3	13.6	45.3	9.6	29.6	35.5	75.3
Neighborhood type (%)								
Stable middle-class	4.9	2.8	13.7	12.6	12.5	12.6	2.6	4.5
Gentrifying yuppie	13.3	8.3	43.6	35.6	9.0	7.3	8.1	7.4
Transitional working-class	—	0.5	31.7	36.3	61.9	66.6	54.5	54.0
Underclass	81.9	88.4	11.0	15.5	16.6	13.4	34.9	34.1
N	308	719	127	237	228	261	148	306

Source: UPFLS

Consistent with evidence on educational attainment presented in chapter 2, Mexicans and Puerto Ricans are the most educationally disadvantaged groups, as evidenced by their low average years of schooling completed (seven and ten years, respectively). Nearly two-thirds of blacks had high school degrees. Black men averaged twelve years of school, which is approximately one year below the average for white men, but well above that obtained by Hispanic men and women. Among women, there were no racial differences in mean years of school completed, but Hispanic women averaged two (Puerto Rican) to five (Mexican) years less education than whites. These educational differentials suggest that blacks might have an advantage in the labor market over Hispanic-origin men, if not whites. However, the welfare participation rates and employment status distributions tell another story, which we investigate in chapters 5 and 6, respectively.

At the time of the survey, almost all (95 percent) of Mexican men were in the labor force (either at work or looking for a job) compared to 91 percent of white, 83 percent of black men, and 84 percent of Puerto Rican men. Consistent with national trends, Puerto Ricans exhibit lower participation rates than Mexicans for both sexes (Tienda 1989; Bean and Tienda 1987). Virtually all Mexican and white men reported that they had worked at some time, but nearly 10 percent of black men and over 5 percent of Puerto Rican men indicated they had never worked by the time of the survey (Tienda and Stier 1991).

Among women, labor force participation rates hovered around 50 percent for all groups except Puerto Ricans, among whom only one in three were in the workforce at the time of the survey. Black women's welfare participation rate in 1987 was comparable to that of Puerto Rican women, yet these groups have very different marital status distributions. Higher shares of Puerto Rican mothers are married compared to black mothers. The lowest welfare participation rates correspond to Mexicans, followed by whites.

Of particular relevance for our study is the distribution of groups by neighborhood type. Blacks are considerably more likely than whites, Mexicans, or Puerto Ricans to reside in underclass neighborhoods (see chapter 2 for more details on this neighborhood typology). Over 80 percent of black parents resided in underclass neighborhoods compared to 14 percent of whites, 15 percent of Mexicans, and 34 percent of Puerto Ricans. On the other hand, approximately half of all whites resided in stable, middle-class, and gentrifying neighborhoods compared to about one-fifth of Mexicans and one-tenth of blacks (13.2 percent) and Puerto Ricans

(11.5 percent). Mexicans are distinguished by their high concentration in transitional working-class neighborhoods, where nearly two-thirds live compared to about one-third of whites and half of all Puerto Ricans.

Tables 3.3 and 3.4 profile the two national urban subsamples (all central-city residents and the urban poor, respectively) using the same characteristics (except neighborhood type) reported in table 3.2 for the UPFLS sample. Chapter 2 showed that minority groups were more highly represented within Chicago's city limit than in the surrounding suburbs. Likewise, nationally the minority composition of central-city populations is higher than the population averages for the United States as a whole. Approximately 18 percent of central-city residents were black, and an additional 12 percent were Hispanic (see table 3.1). The urban poor are even more likely to be black or Hispanic than central-city populations generally, as over half (53 percent) of low-income, central-city residents were so classified. The Chicago and national urban samples of Mexicans differ in one important respect that bears on our main substantive interests, namely, birthplace. Over 80 percent of inner-city Chicago's Mexican-origin residents are foreign-born, compared to 45 percent of their national urban counterparts. This difference is important because immigrants are more likely to be married and to be employed than their U.S.-born counterparts, and they are less likely to rely on means-tested income transfers for support. Thus, many differences between the Chicago and urban national samples of Mexicans can be traced to their birthplace rather than their ethnicity per se. We amplify this point when appropriate throughout the text.

Compared to the Chicago sample (see table 3.2), the national urban sample of parents differs in marital status and educational attainment, as well as in employment and welfare participation rates. There is little age differentiation between the Chicago and national samples (owing to the age restrictions imposed on the NSFH sample we analyze). However, Chicago parents averaged slightly lower levels of education, and this has important implications for poverty and economic well-being. High school graduation rates for blacks were about 55 percent for UPFLS respondents compared to 80 percent for urban parents nationally. Among whites, only 64 and 67 percent, respectively, of men and women who resided in Chicago's inner city had graduated from high school compared to over 90 percent of all urban white parents. Nationwide, about 64 percent of Mexican-origin men who lived in central cities graduated from high school, compared to less than 40 percent of Chicago's Mexican-origin men. For women the comparable figures are 38 and 43 percent, respectively, for residents of Chicago inner-city neighborhoods and other urban centers. Most likely, this reflects nativity differences between the Chicago sample and the national urban sample. That is, foreign-born women are disproportionately represented among Chicago's low-income neighborhoods, and Mexican immigrants average fewer years of education than their native-born counterparts (Greenwood and Tienda 1998).

Chicago parents also were less likely to be married in 1987 (table 3.2) than their national urban counterparts (table 3.3). The higher levels of economic disadvantage among Chicago inner-city parents relative to urban parents nationally are evident in the lower employment rates and higher welfare participation rates, but noteworthy racial and ethnic differences emerge for both samples. Among all urban men, over 90 percent of Mexicans and whites were employed at the time of the survey, compared to only 83 and 77 percent of blacks and Puerto Ricans, respectively. Employment rates of Mexican and Puerto Rican men living in inner-city Chicago were similar to their own-group national averages, but Chicago's black and white men were less likely to hold a job at the time of the survey than their national urban racial counterparts.

For all groups except Mexicans, welfare participation rates were uniformly higher in Chicago compared to urban areas nationally. At the time of the survey, over half of black and Puerto Rican women were on public aid in Chicago, compared to between one-quarter and one-third of black and Puerto Rican women nationally. Again, the higher welfare participation rate of U.S. urban parents compared to Chicago's Mexican-origin parents reflects sample differences in the nativity composition of the population. Although welfare participation rates of Puerto Ricans are not measured reliably because of the small national urban sample, the similarities with the Chicago inner-city sample are striking.

The low-income urban sample profiled in table 3.4 shares many similarities with Chicago's inner city, but also several notable differences. High school graduation rates are higher for NSFH low-income parents than among Chicago's inner-city residents, but slightly lower than for urban parents nationally. This generalization obtains for all demographic groups.[11] Furthermore, low-income urban parents were less likely to be married than either all urban parents or Chicago inner-city parents. For example, only 16 percent of low-income black women (table 3.4) were married at the time of the survey compared to 36 percent of all urban

11. Cells with less than twenty observations are considered too small for reliable analyses. Therefore, we do not separately analyze low-income Puerto Rican men.

Table 3.3 Sample Characteristics by Ethnicity and Sex: Urban Residents, National Survey of Families and Households (Means or Percentages)

	Blacks		Whites		Mexicans		Puerto Ricans	
	Men	Women	Men	Women	Men	Women	Men	Women
Age (years)	33.6	31.8	34.9	33.4	32.3	32.0	34.1	31.6
(s.d.)	(6.3)	(6.7)	(5.6)	(6.6)	(5.8)	(6.6)	(4.7)	(7.4)
Education								
H.S. graduate (%)	79.6	79.7	91.8	90.2	63.6	43.1	70.4	39.2
Education (years)	12.7	12.5	13.8	13.3	10.9	9.6	11.7	9.9
(s.d.)	(2.5)	(1.9)	(2.6)	(2.4)	(3.6)	(3.5)	(3.6)	(3.3)
Marital Status (%)								
Currently married	60.8	36.5	87.3	78.5	80.8	70.6	52.1	39.8
Currently sep/div	21.3	26.7	10.0	17.3	9.2	21.3	27.8	41.3
Never married	17.9	36.8	2.7	4.2	10.0	8.1	20.1	18.9
Current Employment Status (%)								
Employed	82.5	62.3	94.0	64.8	93.2	48.5	77.1	27.7
Unemployed	11.8	10.2	2.2	2.3	5.1	9.5	14.5	5.1
Out of labor force	5.7	27.5	3.7	32.8	1.7	42.0	8.5	67.1
Welfare Participation (%)								
Currently on aid	3.9	27.5	1.7	7.4	2.9	16.4	11.3	34.0
N	213	529	613	1095	101	201	20	62

Source: NSFH, Urban Parent Subsample

Table 3.4 Sample Characteristics by Ethnicity and Sex: Low-Income Urban Sample, National Survey of Families and Households (Mean or Percentages)

	Blacks		Whites		Mexicans		Puerto Ricans	
	Men	Women	Men	Women	Men	Women	Men	Women
Age (years)	32.3	29.8	32.4	30.0	31.5	32.1	—[a]	31.6
(s.d.)	(6.2)	(6.4)	(5.6)	(7.1)	(5.4)	(7.1)		(7.2)
Education								
H.S. graduation	73.9	75.0	79.4	75.0	53.3	23.8		30.8
Education (years)	12.0	12.1	12.3	11.9	10.4	8.3		9.2
(s.d.)	(2.9)	(1.5)	(2.2)	(2.0)	(2.8)	(3.3)		(3.3)
Marital Status (%)								
Currently married	69.6	15.9	88.9	47.7	96.6	58.7		23.1
Currently sep/divorced	4.3	30.7	3.2	42.4	3.4	28.6		53.8
Never married	26.1	53.4	7.9	9.8	0	12.7		23.1
Current Employment Status (%)								
Employed	66.7	43.7	74.6	45.6	89.7	37.1		14.3
Unemployed	19.0	19.5	6.8	6.4	6.9	9.7		7.1
Out of labor force	14.3	36.8	18.6	48.0	3.4	53.2		78.6
Welfare Participation (%)								
Currently on aid	4.3	54.5	12.5	32.3	6.7	31.7		53.8
N	39	218	57	243	34	107	7	32

Source: NSFH, Urban Parent Low-Income Subsample
[a] Sample too small for reliable analysis.

black women (table 3.3) and 23 percent of inner-city black women (table 3.2). The marital status distribution of low-income white women was similar to that of Chicago mothers: approximately half of both samples were married at the time of the interview. Mexican-origin women were most likely to be married of all groups, even if they were poor, but especially if they were foreign-born. About three-fifths of Mexican-origin women residing in central cities were married, but among those residing in Chicago's inner city, nearly 70 percent were married.

Employment rates of low-income urban parents are lower, and unemployment rates are higher than the rates for all urban parents. Among urban men who resided in low-income households, only two-thirds of blacks and three-fourths of whites were employed at the time of the survey. Nearly 90 percent of similarly situated Mexican-origin urban men were employed, which is similar to the employment rate of all urban Mexican-origin men and Chicago Mexicans. Chapter 6 examines these employment differentials and correlates them in great detail. Not surprisingly, welfare participation rates of low-income urban parents were higher than those of their nonpoor urban counterparts with two notable differences vis-à-vis inner-city parents: welfare participation rates of urban poor white and Mexican women were roughly identical, and higher than the welfare participation rates of their inner-city counterparts. Also, among urban black women, the welfare participation rates of the poor were twice as high as the rates of all urban women, but roughly similar to those of inner-city black women.

On balance, these sociodemographic profiles of the UPFLS, the NSFH, and its low-income subsample reveal notable similarities and differences among black, white, Mexican, and Puerto-Rican parents. In general, the urban poor are less likely to be employed and more likely to receive transfer payments at the time of the survey than all urban parents. Similarly, marriage rates are generally higher among the national urban sample than the low-income substratum or the inner-city sample, with the exception of Mexicans. Blacks and Puerto Ricans are the most economically disadvantaged, as evident by their low employment and high unemployment rates, irrespective of the sample compared, and this is so despite their relatively lower educational attainment. This difference is germane to our dominant theme about the relative importance of color and opportunity in determining life chances and economic well-being, which we examine in subsequent chapters.

Conclusion

Two features of the UPFLS make it particularly appealing for our analytical goals. First, the focus on inner-city neighborhoods permits an in-depth portrayal of urban life in low-income neighborhoods, which can be compared and contrasted with studies based on low-income populations. This comparison sharpens the distinction between poor people and poor places, but it is important to remain mindful of the limits of generalizability in the comparisons. Second, the UPFLS is especially attractive for examining racial and ethnic diversity in the way early experiences with poverty affect work, welfare, and family life among inner-city populations, because Chicago is the only large urban center with substantial concentrations of *both* Mexicans and Puerto Ricans, the two largest Hispanic-origin groups. Thus, the Chicago case study permits us to explore the color and opportunity theme in ethnic as well as racial terms, which is a valuable extension of the poverty and underclass debates that occurred during the 1980s and early 1990s. A detailed exploration of the family, welfare, and work behavior of Mexicans and Puerto Ricans is important because the mechanisms driving poverty differ for these groups (i.e., low wages versus lack of work) and because, relative to blacks and whites, much less is known about family dynamics and welfare participation among Hispanics.

CHAPTER FOUR

Family Matters: Turning Points from Orientation to Procreation

Do I think they're given a fair chance to succeed? I don't think that's true for most people, no. Well, I think the size of the families, uh . . . for one thing. Then the composition of the families . . . I think most are single families, with large . . . mainly large families. So I think that when you have large families . . . everybody doesn't get . . . equal opportunity, so to speak. Because you can't afford to give 'em that kind of opportunity. So what happens is one or two maybe make it . . . or sometimes none of 'em make it. And um . . . I think you get into then what is called kind a like a ghetto mentality . . . you can't see yourself gettin' out . . . and that's further complicated by the things you do. You start . . . and the girls start to have babies and then they . . . they're locked into that situation. And they aren't able to um . . . get married or better their situation because of what they've done. The boys think, OK, they want the same type of things that others have and then they get into gettin' those things um . . . illegally. And when that happens, you kind of lock yourself in . . . to poverty.

Felicia, a divorced black mother of four who has never been on welfare, portrays the vicious cycle of poverty in which many inner-city residents are trapped. A schoolteacher by profession, she witnessed firsthand how many urban youth begin life in disadvantage. Their parents are poor and often cannot satisfy even basic needs; many are reared in lone-parent families or with no parents at all. Lacking adequate parental supervision and economic resources, adolescents turn to similarly situated peers for support and guidance, often with devastating results. The peer culture of disadvantage reproduces high rates of teenage childbearing and school failure. In turn, youth who lack basic skills are ill-equipped for the labor market and doomed to unstable work careers, prolonged joblessness, welfare dependence, and limited marriage prospects.

The scenario described by Felicia and paraphrased above as a series of stylized generalizations based on social science research about the consequences of growing up poor presumes that poor families reproduce themselves by transferring poverty intergenerationally. This controversial idea is buttressed by a voluminous sociological literature showing how the family of orientation and the context in which children are reared circumscribe life chances and material well-being in adulthood (Duncan and Brooks-Gunn 1997).

This chapter considers how the early experiences of youth shape alternative pathways from their family of orientation to their family of procreation. Our main goal is to identify race and ethnic variation in pathways to adult family life, and to compare the experiences of inner-city residents to those of urban parents nationally and to those of poor urban parents. We focus on the transition to family life because it is a crucial turning point that has enduring effects on individual socioeconomic prospects, especially the likelihood of remaining in poverty or becoming poor as adults. In particular, we ask whether there are ghetto-specific, group-specific, or context-specific differences in the timing and sequencing of marriage and childbirth. By focusing initially on the family of orientation, we reveal how early experiences with poverty accumulate and reinforce trajectories of disadvantage throughout the life course.

To begin, we profile the family background of the three samples to illustrate racial and ethnic differences in early experiences with poverty and social disadvantage. Subsequently we examine the incidence and determinants of premature school withdrawal and the timing of childbirth, two pivotal events that have prolonged consequences for economic well-being. Third, we describe the alternative pathways to family life, based on the timing and sequencing of marriage and births, and relate these to early experiences with poverty. We argue that premature entry to family roles, particularly when births precede marriage, perpetuates social and economic disadvantages from the family of orientation to the family of procreation. We conclude the chapter by examining income and welfare consequences associated with alternative pathways to adult family life.

Both in Chicago and in urban areas nationally, we document large racial and ethnic differences in family structure and early experiences with poverty, finding that blacks are more likely than whites or Hispanics to have been reared by a lone parent and to have been exposed to welfare as a child. That Chicago parents are more likely than all urban parents to have been exposed to poverty during childhood is socially significant because it influences the likelihood of high school completion and pathways to family life. On balance, our analyses reveal that childhood experiences with poverty perpetuate further disadvantages over the life course by truncating educational careers and facilitating entry to nonmarital childbearing. Although controlling for group differences in family structure and early exposure to poverty largely diminishes racial and ethnic differences in high school completion and pathways to family life, differences in educational attainment between Hispanics and whites persist.

We do find significant racial differences in family formation patterns, with black women significantly more likely than whites or Hispanics to enter family life via birth. This pattern obtains among all urban women, and especially the urban poor, but nonmarital fertility is more prevalent in Chicago, and it occurs at earlier ages compared to all urban mothers. Chicago's black mothers exhibit somewhat lower rates of marriage than poor urban mothers nationally, but they exhibit a strikingly similar pattern of family formation via premarital births. Thus, we conclude that inner-city mothers do not behave differently from poor urban mothers in general, and that both race and early experiences with poverty independently influence the likelihood of entering family life via birth or marriage. Given racial and ethnic differences in pathways to family life, the income and welfare participation consequences are fairly uniform across groups.

The Family of Orientation

Families provide the primary social context that shapes lives of youth. Not only are values and aspirations established and reinforced, but youngsters witness adult behaviors that they emulate as they mature—often prematurely. Several aspects of the family of orientation mold young adults' lives and determine their future opportunities. First and foremost, family structure, and in particular the absence of a father, is associated with several forms of problem behavior, including premature high school withdrawal, teenage childbearing, and early encounters with the criminal justice system (Wu 1996; McLanahan and Sandefur 1994). Family structure also affects children's cognitive development, success in school, and ultimately their adult life scripts (Brooks-Gunn, Duncan, and Aber 1997).

Second, a family's material resources have decisive, though not irreversible, consequences for children's development and transition to adulthood. Children reared in poverty often reside in poor neighborhoods, which affect their exposure to non-normative behaviors and role models. Also, and depending on the type of neighborhood in which poor children are raised, the quality of the schools they attend and the safety of their physical environments influence whether they develop into healthy and capable adults or reproduce the material culture of disadvantage in which they were reared.

Finally, parents' education is a crucial resource for children because it sets a floor below which children are unlikely to fall in their own scholastic achievement (Mare 1995). Parents with a high school education, even if poor, can appreciate the value of education for labor market success and are highly likely to encourage their children to complete at least secondary school (Coleman 1990; Sewell and Hauser 1975; Mare 1995; Ahituv and Tienda 1996). Educational attainment is the key to intergenerational upward mobility, yet the weight of prior ascription slows the pace of educational progress for successive generations, and particularly minority groups.

We begin our empirical inquiry of pathways to adult family life by comparing the family circumstances in which UPFLS respondents were reared, drawing selective comparisons with their national urban counterparts to ascertain whether and how inner-city parents differ from urban parents in general and from the urban poor more specifically. We seek to identify group-specific and place-specific differences in early life circumstances that may help explain variation in adult disadvantages. Table 4.1, which presents selected characteristics of the family of origin for the three comparison samples, warrants several generalizations along these lines. First, blacks and Puerto Ricans were more likely than either Mexicans or whites to have been reared in single-parent families, but the relative shares differ by context. Both among urban parents nationally and in Chicago's inner city, between 40 (women) and 50 (men) percent of black parents were reared in two-parent families, but only one in three urban poor mothers were so reared. By this measure, Chicago's inner-city parents are not worse off than their national urban counterparts, and they are better off, on average, than poor urban parents nationally. Similarly, higher shares of Puerto Rican Chicago parents were raised by both a mother and father compared to all urban Puerto Ricans. Among urban parents nationally, approximately 70 percent of Mexicans and whites were reared in two-parent families, but only half of their urban poor counterparts enjoyed the benefit of both a mother and a father when growing up. Thus, based on this dimension of the family of orientation, Chicago's inner-city parents were not clearly more disadvantaged than their national urban counterparts, especially urban poor parents.

Second, across samples, blacks and Puerto Ricans were more likely than either whites or Mexicans to have been reared in poverty, which we measure based on exposure to welfare during childhood and adolescence. However, early life experiences with poverty were far more common among Chicago parents than urban parents nationally, with the notable exception of Mexicans. For example among Chicago blacks, whose rates

Table 4.1 Selected Characteristics of Families of Orientation: Parents Residing in Chicago's Inner City and U.S. Cities (Means or Percentages)

	Blacks		Whites		Mexicans		Puerto Ricans	
	Men	Women	Men	Women	Men	Women	Men	Women
Chicago								
Both parents to age 14	50.8	45.4	72.5	66.1	83.2	79.0	55.2	56.3
Parents ever on welfare	30.0	40.7	5.7	17.0	2.2	6.2	30.7	37.7
# Siblings	5.3	5.8	3.5	3.8	6.8	6.5	6.3	6.4
(s.d.)	(3.5)	(3.4)	(2.7)	(2.7)	(2.9)	(3.0)	(3.5)	(3.7)
Sisters unwed mom	46.8	57.1	14.0	21.1	12.8	13.9	28.7	38.7
Brothers unwed dad	47.7	47.5	12.3	13.7	10.0	19.4	34.3	38.6
N	308	719	127	237	228	261	148	306
U.S. Cities								
Both parents to age 14	52.0	40.1	71.4	71.0	72.3	65.2	—[a]	43.0
Parents ever on welfare	17.4	27.1	9.8	8.1	14.6	16.0	—	35.7
# Siblings	4.6	5.1	3.1	3.0	6.2	6.1	—	5.4
(s.d.)	(3.2)	(3.2)	(2.2)	(2.2)	(3.3)	(3.3)		(3.9)
N	213	529	613	1,095	101	201		62
U.S. Poor								
Both parents to age 14	—[a]	32.4	—[a]	56.6	—[a]	54.3	—[a]	—[a]
Parents ever on welfare	—	38.0	—	20.0	—	18.6	—	—
# Siblings	—	5.1	—	3.0	—	6.1	—	—
(s.d.)		(3.2)		(2.1)		(3.6)		
N		218		243		107		

Source: UPFLS and NSFH, Urban Parent and Low-Income Subsamples
[a] Sample size too small for reliable estimates.

of early exposure to welfare are the highest of all groups compared, 30 to 40 percent of men and women, respectively, received public assistance when growing up, but only 17 to 27 percent (men and women, respectively) of their national urban racial counterparts were on welfare grants as youth. The shares of Chicago black mothers and urban poor black mothers reared in welfare-reliant homes were virtually identical. Also, nearly one-third of Puerto Rican parents were exposed to welfare during their childhood, and this share was similar across samples. Chicago's white parents were more likely to have been reared in poverty than white urban poor parents nationally. This reflects the Chicago sample-selection criteria, which deliberately overrepresented parents from poor places.

As we show in chapter 5, for most urban white parents, poverty spells are relatively short, often stemming from changes in family circumstances. But for blacks, particularly those residing in high-poverty neighborhoods, poverty and welfare spells are often chronic and intergenerational. In fact, the share of white mothers exposed to welfare during their childhood was similar between the Chicago and the urban poor subsamples. The low rates of welfare exposure among Chicago's Mexican-origin parents reflect their immigrant status. Because public assistance is not available in Mexico, foreign-born parents cannot have been exposed to welfare during their childhood. This is not so for the national sample of Mexican parents, among whom the native-born predominate (only 40 percent of the national sample is foreign-born compared to 80 percent of Chicago men and women). Thus, rates of exposure to welfare are higher among Mexicans in the urban national sample compared to those for Chicago's largely immigrant subsample, reflecting differences in eligibility and propensity to receive public aid.

Third, minority parents generally were reared with more siblings than their white counterparts, whose sibships average 3.6 among Chicago residents and about 3 among all urban parents. Hispanics were reared in larger families than either blacks or whites, and this aspect of early family life is consistent across samples. The average sibship size for Mexican-origin parents was 6.6 children for those residing in Chicago (mainly recent immigrants); it was slightly lower—6.1—for Mexican urban parents nationally, including the poor urban substratum. Puerto Rican parents were reared in similarly large families, with sibships averaging 6.4 among Chicago respondents and 5.4 among urban Puerto Ricans nationally. Sibship size is important for understanding early development experiences, not only because it indicates competition for material resources and parents' time, but also because siblings' behavior, like that of parents and other adults, serves as a frame of reference for younger children.

In this vein, it is relevant that nearly half of the siblings of Chicago inner-city parents experienced a nonmarital birth, compared to one-third of Puerto Rican parents. In Chicago, only 21 percent of white mothers and 14 percent of Mexican mothers had sisters who bore children out of wedlock; 14 and 19 percent of these women, respectively, had brothers who fathered children out of wedlock. If early exposure to nonmarital childbearing influences the likelihood that younger siblings will themselves become unmarried parents, then we should expect that the incidence of nonmarital childbearing will be highest among blacks and Puerto Ricans, which we demonstrate below. However, we are not able to relate siblings' nonmarital childbearing directly to respondents' risks of becoming unmarried parents because the UPFLS does not indicate whether respondents' siblings were in fact older and actually influenced their behavior. Nevertheless, based on the answers reported in table 4.1, these patterns *do* reflect environments where such behavior was common, if not pervasive. Lacking comparable information for the urban subsamples, we cannot say whether the observed differentials represent group-specific or place-specific differentials. What we can say is that the prevalence of premarital births within families provides yet another indicator of early disadvantage that is rooted in material deprivation and leads to problem behavior during adolescence that reverberates through the family of procreation.

Finally, some gender differences exist in family background among racial and ethnic groups, with the notable exception of whites. Women residing in Chicago's inner city tend to come from more disadvantaged backgrounds than their male counterparts, as evident in the shares reared by lone parents and in welfare-reliant families. Also, women were more likely to have sisters who became single mothers themselves than men were to have brothers who fathered children out of wedlock. Because these differences are less acute among urban parents nationally, one might surmise either that residence in poor neighborhoods increases poor women's vulnerability to deleterious outcomes, or that the selection processes that sort poor women into high-poverty, inner-city neighborhoods render them more vulnerable to these adverse outcomes than is true for men. A third possibility is that men underreport children fathered out of wedlock either because they do not know about them or refuse to acknowledge them formally, even if they do so informally (Stier and Tienda 1993).

In summary, there are small place-specific differences in respondents' family structure, but relatively large racial and ethnic differences. However, table 4.1 shows substantial differences in respondent's early poverty experiences across settings. Chicago residents are more likely than their national, urban, racial/ethnic counterparts, but just as likely as their poor urban counterparts, to have been reared in poor families, as measured by their early exposure to welfare. These group-specific parallels across settings lend support to claims that adult economic hardships are at least partly transmitted intergenerationally between the family of orientation and the family of procreation.

Parental absence and material deprivation reinforce the intergenerational transmission of poverty through several mechanisms, two of which our survey data enables us to investigate empirically, namely, failure to complete high school and premature entry into adult family roles. Both outcomes are directly related to family structure, which usually means inadequate supervision of adolescents, especially when many siblings compete for the time of lone parents. The challenges confronted by lone parents in supervising their children are aggravated by material constraints imposed in the context of concentrated urban poverty. Residing in poor neighborhoods with inadequate schools and weak community infrastructure to occupy youth during the off-school hours also reinforces the deleterious effects of early experiences with family poverty and weak parental control. Because the interruption of high school education is a critical turning point in the life course, we investigate how early disadvantages and economic hardships experienced at home influence the risk of high school noncompletion and adolescent childbearing for minority and nonminority youth residing in Chicago and U.S. central cities.

Parental Supervision and Pivotal Problem Behavior

Lack of parental supervision is one reason why adolescents digress from "normative" developmental paths and instead engage in problem behavior (McLanahan and Sandefur 1994; Hogan and Kitagawa 1985). McLanahan and Sandefur (1994) found significant differences between single- and two-parent families in the effectiveness of parental control over children. Quite simply, in most circumstances, two parents have more time to invest in children than do single parents, particularly if one parent devotes little or no time to the labor market. They attribute this finding to the lower level of "parental resources" in single-parent families compared to two-parent families. Not surprisingly, McLanahan and Sandefur (1994) also found that the level of parental resources and the effectiveness of parental control directly affect the likelihood that children will complete high school and bear a child out of wedlock. Therefore, we begin by examining group variation in parental supervision in order to identify differences in vulnerability to adolescent childbearing and failure to complete high school—two pivotal life-course events.[1]

Parental supervision practices reported in table 4.2 differ mainly by sex, and to a much lesser extent by minority-group status. Among all groups compared, girls were much more closely supervised than boys. For example, sex differences in parental control over their child's whereabouts range between 10 percent (whites and Mexicans) to 18 percent (blacks and Puerto Ricans). Similar sex differences obtain for other aspects of parental supervision, including friends and time spent alone. Sex differences in parental supervision were largest for decisions about friendship associations and smallest for amount of unsupervised time. Sex differences in the amount of parental supervision reaffirm the common

1. Because the NSFH lacks comparable information, we restrict analyses of parental supervision to Chicago inner-city parents.

Table 4.2 Ethnic and Sex Variation in Parental Supervision during Adolescence (Percent Responding Very Much)

Domain of Parental Supervision	Blacks		Whites		Mexicans		Puerto Ricans	
	Men	Women	Men	Women	Men	Women	Men	Women
Where went as teen	64.8	81.9	53.1	63.3	67.1	78.5	64.8	82.9
Who saw as teen	43.3	66.6	35.1	52.0	43.3	69.6	49.8	73.6
Curfew time as teen	73.2	86.3	66.1	79.9	69.9	80.0	70.5	84.2
Almost always left alone after school	25.3	15.9	35.8	25.4	21.8	12.0	19.2	15.9
N	308	719	127	237	338	261	148	306

Source: UPFLS

perceptions that boys and girls face different types and levels of risks when left unsupervised. To wit, unsupervised girls can and often do become pregnant.

Racial and ethnic differences in levels of parental supervision are also evident, although less pronounced. White boys and girls reported the lowest level of parental control during adolescence, but there were minor differences in parental supervision among blacks, Mexicans, and Puerto Ricans. Only 63 percent of white girls reported that their parents supervised their whereabouts, compared to about 80 percent of minority girls. For boys, the comparable figures were 53 percent for whites compared to approximately 65 percent for blacks and Hispanics. White boys and girls also were more likely to be left alone at home—36 and 25 percent, respectively—compared to only 12 to 15 percent of minority girls and 20 to 25 percent of the boys. Most respondents claimed they had curfew times during adolescence, with whites less likely than minority youth to experience time limits on their evening social activities. Unfortunately, our data do not contain information about the actual curfews, nor about the ages at which curfews were imposed.

In light of the rather large differences in family structure reported in table 4.1, racial and ethnic variation in parental practices seem small by comparison. With the UPFLS (especially because of the inability to compare it with the national urban population), it is hard to assess whether similarities in parental supervision among minorities are related to residence in poor neighborhoods, which may impel parents to monitor their children more closely, or whether minority parents in general keep closer tabs on their children than nonminority parents. Overall, the tabular findings indicate that white parents seem to exercise weaker control over their adolescent children, but this may simply reflect the fact that they live in safer environments. Chicago neighborhoods inhabited by poor whites are nowhere near as poor or dangerous as those inhabited by minorities (Tienda and Stier 1991; see also chapter 3).

To the extent that problem behavior is governed primarily by group-specific differences in parental supervision, the evidence presented in table 4.2 would lead one to expect relatively small racial and ethnic differences in adolescent problem behavior. But if group differences in opportunities for deviant behavior are largely dictated by early life socialization experiences and normative environments conducive to non-normative behavior, then we would expect larger racial and ethnic differences in adolescent problem behavior than suggested by table 4.2. This follows because minority families are poorer, and, at least for blacks and Puerto Ricans, youth are more likely to have been reared in lone-parent families, with all of their attendant fragilities.

Because of their profound implications for adult well-being, we consider two pivotal events that link economic disadvantages intergenerationally, namely, premature departure from school and adolescent childbearing. Dropping out of high school and early childbearing limit young adults' ability to accumulate economic and social resources over the life course. Failure to complete school has major implications for future employment opportunities and thus for the risk of welfare participation in adulthood. Now more than in the past, the jobs available to poorly educated workers do not pay family wages. In *Growing Up with a Single Parent,* McLanahan and Sandefur (1994) present strong and compelling evidence that children reared in single-parent families are more likely than those with both parents at home to drop out of high school. Thus this early life-course decision is both directly and indirectly related to adult poverty risks.

McLanahan and Sandefur (1994) also show that girls reared in welfare-reliant homes face higher risks of a premarital birth. Although there is some debate about the exact consequences of adolescent childbearing, there is general agreement that teen parenthood compromises the life chances of both mothers and their children. Bearing a child out of wedlock affects the probability of marrying at all, and also increases the likelihood of marital dissolution among those who eventually do wed (Miller and Moore 1990; Cutright and Smith 1988; Nathanson and Kim 1989). These generalizations hold despite racial and ethnic differences in the likelihood of teens marrying the father of their child (Testa et al. 1989). Teen parenthood is thus a pivotal point in the socioeconomic life course of poor women: it is related to dropping out of high school (D. Anderson 1993;

Forste and Tienda 1992); it affects labor market options (Ahituv, Tienda, and Tsay 1998); and it is one of the major paths to welfare dependence (Wu 1996; McLanahan and Sandefur 1994).

In contrast to the relatively small group differences in problem behavior reported in table 4.2, appreciable racial and ethnic differences exist in adolescent problem behavior in our Chicago sample, as well as in the national urban and poor urban subsamples. Table 4.3 presents the distribution of these two problem behaviors for all three samples. Beginning with educational attainment, the top panel of table 4.3 shows that Mexicans and Puerto Ricans exhibit the highest dropout rates, as the majority (around 80 percent) did not graduate from high school. By comparison, half of black respondents in Chicago and about 40 percent of whites did not complete high school by the age of nineteen. As shown in the second and third panels, racial and ethnic differences for Chicago parallel those among urban parents nationally, although average dropout rates among Mexicans nationally are lower than those observed among Chicago's largely immigrant population. Among the urban poor, virtually all Mexican-origin women (90 percent) interrupted their education before graduating from high school. However, low education is a general characteristic of U.S. Hispanics rather than a unique characteristic of inner-city residents, although the low attainment levels are compounded by material deprivation.

Racial and ethnic groups also differ appreciably in their rate of teen childbearing. Black females (but also black men, relative to other men) followed by Puerto Ricans have the highest incidence of teen parenthood. One of every three black women and one-fifth of all Puerto Rican women residing in Chicago's inner city experienced a premarital birth as a teenager. Mexicans' rate of teen childbirth also is relatively high, but it is related to their early age of marriage. Only 10 percent of Chicago Mexican-origin women reported having their first child out of wedlock, which is similar to the rate for white women, but poor Mexican-origin women nationally had a higher risk of a nonmarital teen birth. In comparison to Chicago, the incidence of nonmarital and teen childbearing among all central-city dwellers is appreciably lower. Only one-quarter of black urban women (compared to 40 percent in Chicago) and 4 percent of whites (versus 9.4 percent in Chicago) experienced a premarital birth.

Among the urban poor, out-of-wedlock teen motherhood is as prevalent as in Chicago's poor, inner-city neighborhoods, and racial and ethnic differences run parallel. That is, poor urban blacks are the most likely to bear children out of wedlock. Specifically, about one-quarter of all urban black mothers experienced a premarital birth as teens compared to only

Table 4.3 Selected Indicators of Problem Behavior during Adolescence (Means or Percentages)

	Blacks		Whites		Mexicans		Puerto Ricans	
	Men	Women	Men	Women	Men	Women	Men	Women
Chicago								
Nongraduate by age 19	54.0	54.6	41.6	42.3	80.9	76.2	80.4	78.2
First child as teen	10.7	40.0	.9	15.8	3.7	19.5	2.8	30.3
First child as unwed teen	9.5	34.5	.9	9.4	.6	10.0	2.8	19.7
N	308	719	127	237	228	261	148	306
Urban United States								
Nongraduate by age 19	42.8	39.9	24.6	20.9	64.1	71.4	—	80.0
First child as teen	5.0	30.3	2.8	10.8	8.2	21.2	—	35.5
First child as unwed teen	4.3	25.6	2.0	4.2	4.1	11.7	—	13.8
N	213	529	613	1095	101	201		62
U.S. Poor								
Nongraduate by age 19	—[a]	46.5	—[a]	47.0	—[a]	90.3	—[a]	—[a]
First child as teen	—	42.0	—	27.3	—	25.4	—	—
First child as unwed teen	—	38.6	—	13.6	—	17.5	—	—
N		240		248		113		

Source: UPFLS and NSFH.

[a] Sample size too small for reliable estimates.

4 percent of white women, 12 percent of Mexicans, and 14 percent of Puerto Ricans who resided in central cities during the late 1980s. But among the urban poor, almost 40 percent of black women compared to only 14 percent of whites and 18 percent of Mexican women had a premarital birth. The figures for men show a considerably lower rate of teen parenthood, although the racial differences persist both in Chicago and in the urban United States.

In sum, the tabular analysis reveals large—sometimes striking—racial and ethnic differences in these two behaviors that compromise life options. But it is not clear whether the observed differences reflect group-specific behavior, or whether the four populations compared differ in characteristics that are systematically related to premature high school withdrawal and adolescent childbearing. We hypothesize that the large group differences derive from racial and ethnic variation in family structure and early experiences with poverty and *not* to group-specific behavioral syndromes or to variation in parental supervision. To address this possibility, we turn to multivariate analyses.

Determinants of Premature High School Withdrawal

Figure 4.1 presents racial and ethnic differences in the probability of high school noncompletion for Chicago, U.S. central cities, and low-income urban men and women. The probabilities were derived from a logistic regression analysis predicting high school noncompletion as determined

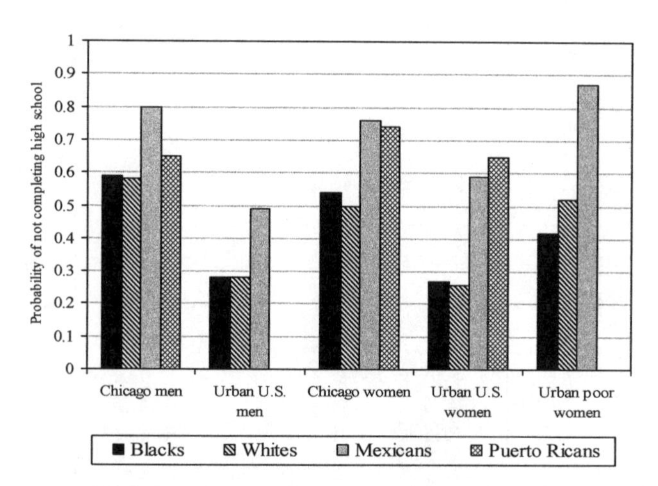

Figure 4.1 Risk of High School Dropout by Race, Ethnicity, and Sex: Chicago and Urban United States

by race and ethnicity, family background, and age-cohorts. Empirical results are presented in appendix table B4.2. The groups' average characteristics (reported in appendix table B4.3) were used to derive the reported probabilities using the procedure detailed in the methodological appendix.[2]

Controlling for group characteristics, the figure shows that racial differences in the risk of not completing high school were virtually eliminated by adjusting the gross rates for family background, sibship size, and age cohort. However, the high rates of high school withdrawal observed for Hispanics persist after controlling for these characteristics. Specifically, the probability that inner-city Mexican and Puerto Rican men drop out of high school is 0.8 compared to 0.6 for Chicago whites and blacks. Similar differences in dropout risks are observed for Chicago women.[3] The probability that inner-city Hispanic women withdraw from high school is 0.75 compared to 0.54 and 0.5, respectively, for black and white women. Similar racial and ethnic differences obtain for the urban national sample. Hispanic men are almost twice as likely to interrupt their secondary education compared to their black and white counterparts, and the differences are slightly higher among urban women. Nationally, Mexican and Puerto Rican urban women are, respectively, 2.2 and 2.5 times more likely to drop out of school compared to black and white women with similar family background characteristics.

There are striking place-specific differences in the risk of high school noncompletion. For all racial and ethnic groups, the probability of dropping out of high school is higher for Chicago residents than for central-city dwellers nationally. Inner-city black and white men are twice as likely as their urban counterparts to interrupt their education (a probability of 0.58 for the former compared to 0.28 for the latter), and the relative risks are similar for Hispanics (a probability of 0.8 in Chicago compared to about 0.45 in the national sample). Similar place differences obtain for women. These persistent differences in the probability of premature high

2. By inserting the mean value of each covariate other than the variable of interest—race and ethnic origin in this instance—we compare the probability of a black man with the average characteristics of the Chicago male sample to a white or Hispanic man in Chicago with the same characteristics. Following with this example, 59 percent of Chicago men were raised by both parents, 35 percent of them had a mother who completed high school, and 59 percent grew up in poor families. Among U.S. men, 68 percent were raised by both parents, and only 12 percent grew up in poverty.

3. Although the magnitude of the difference in the probability is small, it is statistically significant. These estimates are based on a logistic regression.

school withdrawal between residents of poor neighborhoods and those residing in central cities appear to be driven by poverty, not by group membership. Comparisons with urban poor women support this conclusion. The adjusted probabilities of high school noncompletion for urban women are almost identical to those observed for Chicago women, except that black inner-city mothers, the majority of whom reside in ghetto-poverty neighborhoods, have a higher probability of dropping out of high school (0.53 compared with 0.42 for urban poor mothers). This comparison suggests that the intensity of neighborhood poverty does, in fact, intensify the deleterious effects of family poverty. W. J. Wilson (1987) referred to such outcomes as "concentration effects."

Growing up with a single parent does not influence the probability of high school completion for inner-city men, but its detrimental effect on educational attainment is more pronounced for inner-city women and even more so for urban women and men nationally. As figure 4.2 shows, the risk of not completing high school is 11 percent higher for inner-city mothers who grew up with one parent compared to those who were raised by two parents. For urban men and women reared by one parent, the probability of not graduating from high school is 13 and 17 percent higher than for their counterparts who grew up with both parents. The weaker effect of family structure on the educational attainment of inner-city residents may reflect the overpowering influence of other environmental

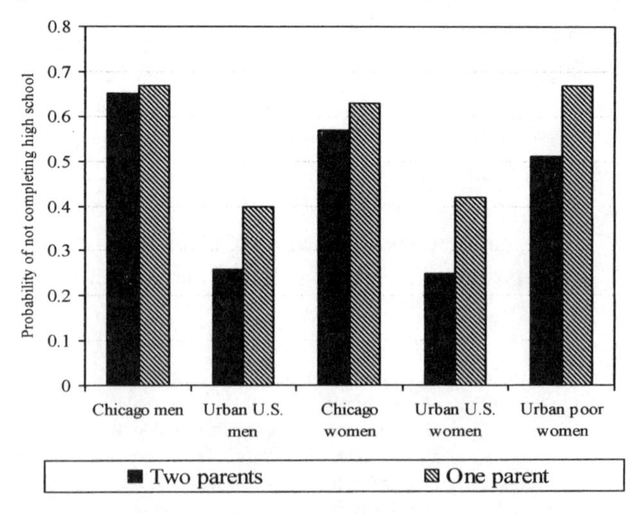

Figure 4.2 Risk of High School Dropout by Family Structure and Sex: Chicago and Urban United States

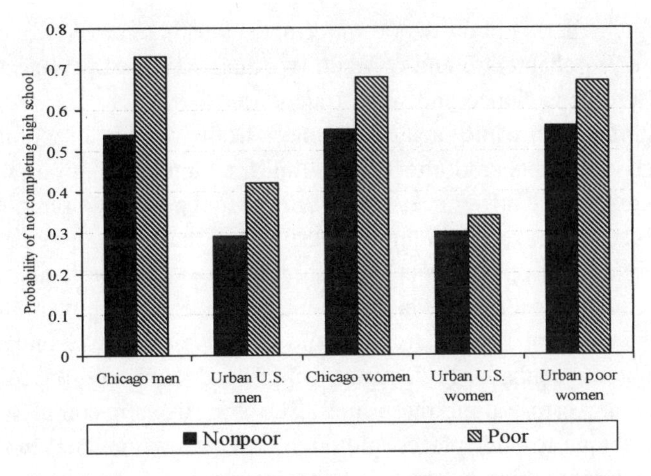

Figure 4.3 Risk of High School Dropout by Family Poverty Status and Sex: Chicago and Urban United States

influences conducive to school failure, which in turn presents as higher *levels* of noncompletion. Not only are many inner-city schools dangerous, but high-poverty neighborhoods often lack the role models to encourage success in school and support normative development. These social forces affect all groups uniformly and, given the high rates of high school failure, powerfully. In W. J. Wilson's (1987) terms, poor, inner-city neighborhoods lack the social buffers needed to prevent social disorganization and its attendant deleterious outcomes for youth.

Although family structure has a relatively weak effect on the educational attainment of inner-city men, family poverty status significantly influences the likelihood of graduation. Growing up in poverty has strong, adverse consequences for educational attainment both among Chicago residents and among the urban poor generally (see figure 4.3). Exposure to welfare as a child increases the probability of Chicago's inner-city men dropping out of high school by 13 percent, and that of inner-city women by 12 percent. Comparable effects obtain for urban men and women nationally. That is, childhood experiences with welfare increase the likelihood that central-city residents drop out of high school by 12 to 14 percent for women and men, respectively, including the urban poor. These figures confirm results of prior studies showing that exposure to poverty during childhood perpetuates disadvantages by truncating educational careers. This finding is robust inasmuch as it obtains for all groups, irrespective of where they live or how poor they become later in life. It remains to be seen whether and how school interruption leads to poverty, a question addressed in chapters 5 and 6, where we analyze the effect of schooling on welfare dependence and work trajectories.

To summarize, while racial differences in the risks of dropping out of high school disappeared after controlling for family background, ethnic- and place-specific differences persist. Because the Chicago sample under-represents low-poverty neighborhoods, and because the survey lacks direct measures of respondents' exposure to poor neighborhoods, it is not possible to assess directly whether living in concentrated poverty accentuates the effects of family structure and early welfare experiences or the likelihood of premature high school withdrawal. However, the results are strongly suggestive along these lines. The low Hispanic completion rates are not unique to poor places, although they are accentuated in contexts of deprivation, which suggests a role for policy.

Analyses also reveal two additional mechanisms through which the family of orientation transmits disadvantage over the life course, namely, family structure and early exposure to material deprivation. The educational consequences of early exposure to welfare are relatively uniform across place, but family structure effects are more pronounced for inner-city women compared to their male counterparts. Taken together, our findings reveal that starting life in disadvantage perpetuates further disadvantages which, in the long run, reproduce negative consequences in respondents' families of procreation.

Inner-city residents understand the importance of education as a requirement for getting ahead in life. When asked what is needed to succeed in Chicago, the majority of participants in the Social Opportunity Survey recognized the importance of formal education. Rebecca, a separated white mother of two who has a high school diploma and has participated continuously in the labor force, put it this way:

> *Get a education . . . get an education. Stay in . . . if you're in school, stay in school. Get all, get all, you know, get your grammar school and your high school, get that all taken care of. 'Cause ah, if you don't get that, I don't think you can get a job. It's harder if you don't have the four years of high school. If you don't have the diploma, it's really hard.*

Truncated educational careers are but one aspect of disadvantage with lasting effects throughout the life course. The transition into family life is another pivotal point in the life course that compromises the life chances of youth reared in poverty. Therefore, next we consider how intergenerational continuities in poverty are transmitted via decisions about the timing and sequencing of births and marriage.

Pathways to Family Life

> *There's a lot of teenager children . . . younger teenagers these days that have babies . . . the young people being parents really is . . . not . . . not ready, they're not ready to have children yet. . . . I wanna put the blame on the parents, you know, but . . . it really takes two, so I don't know where to . . . put all the blame on the parents for not bringing up the children the right way or teaching 'em right, or maybe they're not setting a good example for their children . . .*

Although Rebecca does not elaborate on the consequences of adolescent childbearing, her comments clearly convey that premature entry into adult family roles is a serious problem. Of the two "pathways" to begin family life—through childbirth and through marriage—the latter remains normative, both statistically based on prevalence and socially based on acceptability. These alternative pathways have direct implications for well-being in adult life. Nonmarital childbearing, especially during the teen years, is associated with less-than-average education and a lower propensity ever to marry. Most important, out-of-wedlock childbirth often leads to reliance on public assistance for income support (Forste and Tienda 1992; Furstenberg 1976; Hofferth and Moore 1979; McLaughlin et al. 1988; Menken 1980; Taylor et al. 1990).

The timing of the transition to family life also has critical effects on other life events. Premature transitions into family life (whether via marriage or birth) are related to higher risks of family dissolution, larger families, and, for women, also weaker attachment to the labor market (McLanahan and Booth 1989)—outcomes that are themselves related both to material disadvantage and to group membership. Therefore, in documenting variation in pathways to family life among inner-city residents and the U.S. urban population at large, we emphasize both racial and ethnic differences and early life experiences that are correlated both with group membership and family formation behavior. Seeking evidence of ghetto-specific behavior, we also compare inner-city parents to urban parents nationally.

Figures 4.4 and 4.5 depict the patterns of entry to family life via marriage, respectively, for women and men residing in Chicago and U.S. central cities. The survivor functions presented in the figures are based on life-table techniques discussed in appendix B. The curves indicate the percent of the population that had not yet married at specific ages. Teenage marriage is relatively uncommon. However, the rate of marriage increases after age seventeen for women in Chicago and in U.S. central cities, and approximately two years later for men. The pace of entry to marriage (as evident by the shape of the survival function) is slower among most groups residing in Chicago compared to central-city populations generally. Poor men and women in Chicago are less inclined to marry at all than their urban counterparts nationally.

Group-specific differences in the timing of marriage are similar for men and women, and among all groups, women enter marriage faster than men do. Mexicans in Chicago still exhibit the highest rates of marriage, and they enter family life at earlier ages (at a faster rate) than any other group. This is consistent with the high immigrant composition of Chicago's Mexican population. For the U.S. as a whole, marriage rates of Mexican and white women are virtually identical. The main difference in the rate of family formation via marriage is along racial lines, and more so in the inner city of Chicago than central cities on average. Chicago blacks

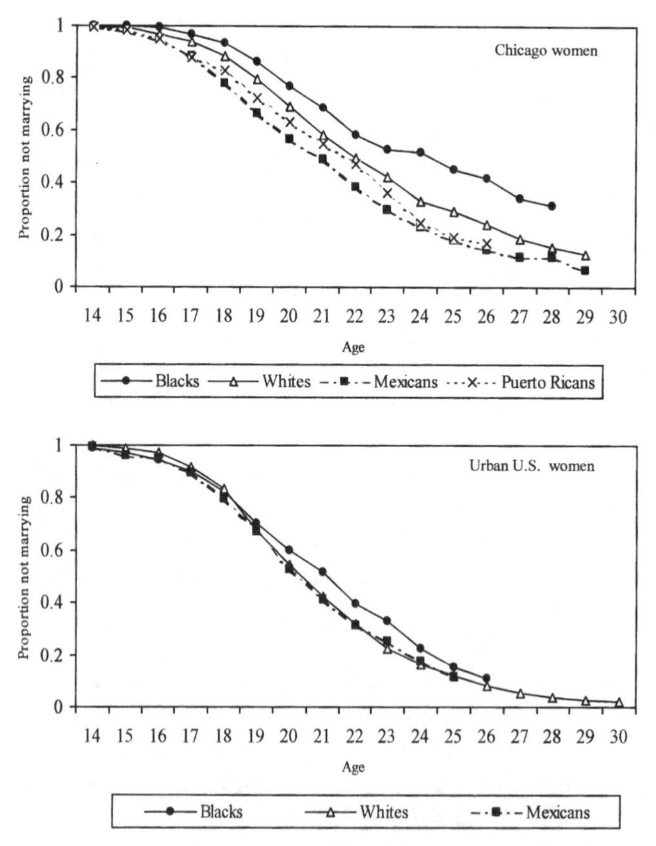

Figure 4.4 Marriage Pathway to Family Life: Chicago and U.S. Urban Women by Race and Ethnicity

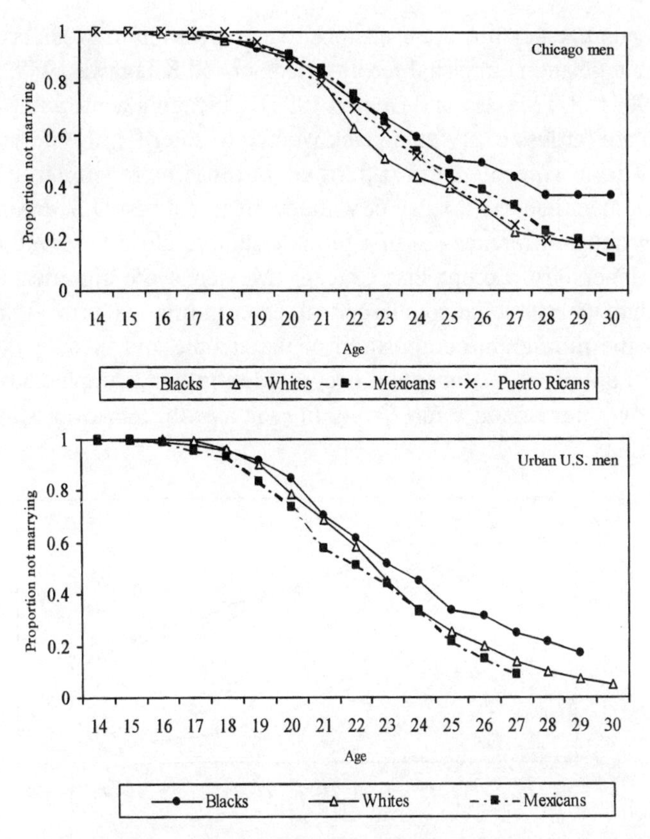

Figure 4.5 Marriage Pathway to Family Life: Chicago and U.S. Urban Men by Race and Ethnicity

exhibit the lowest propensity to marry at any age, but their white and Hispanic counterparts behave similarly (based on the shape of the survivor function) with respect to the timing of marriage. By contrast, racial differences in the timing of marriage were barely discernible in the national sample. This difference in the rate of entry to marriage between inner-city blacks and all urban blacks raises questions about the relative salience of race versus place in marriage behavior. We consider the possibility of ghetto-specific marriage behavior below.

The comparison of the birth pathway into family life unveils a different story. At each age, the probability of entering family life via a premarital birth is appreciably higher in Chicago's inner city than in urban areas nationally, as indicated in figure 4.6 for women in Chicago and U.S. central cities, and in figure 4.7 for men, respectively. This finding is consistent with generalized claims about the positive association between economic disadvantage and nonmarital fertility (Hogan and Kitagawa 1985; Anderson 1990, 1991; Massey and Denton 1993). Chicago's white and Mexican women are far less likely than black women to enter family life via birth. Puerto Rican women's entry pattern via births is more similar to that of black women than to Mexican or white women. Among U.S. urban population groups, differences in first births resemble those observed in Chicago's inner city, except that Chicago Mexicans are significantly less likely than their national counterparts to enter family via birth. Again, this reflects the immigrant composition of the sample and their tendency to marry in the event a conception occurs. Men in both samples have relatively low rates of entry into family life via a birth. However, Chicago's

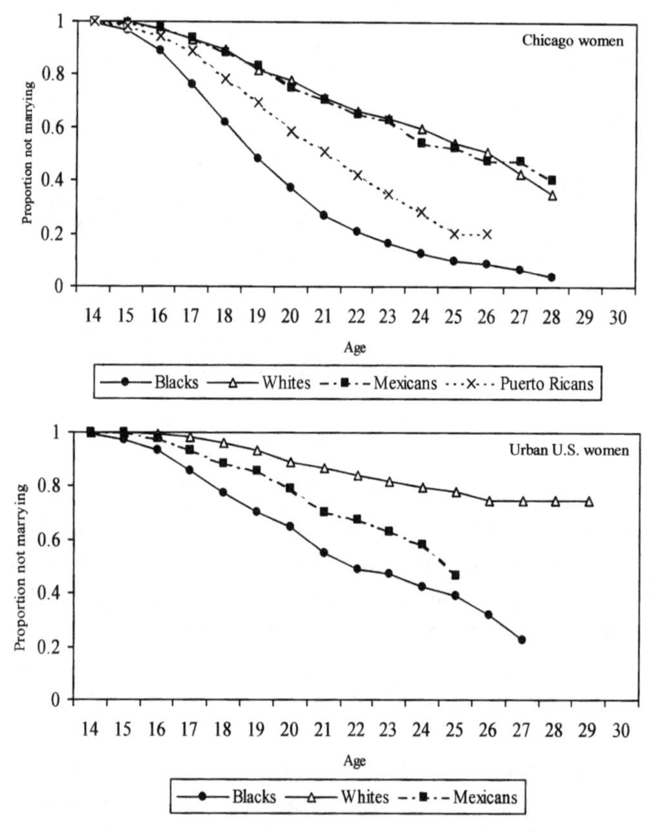

Figure 4.6 Birth Pathways to Family Life: Chicago and U.S. Urban Women by Race and Ethnicity

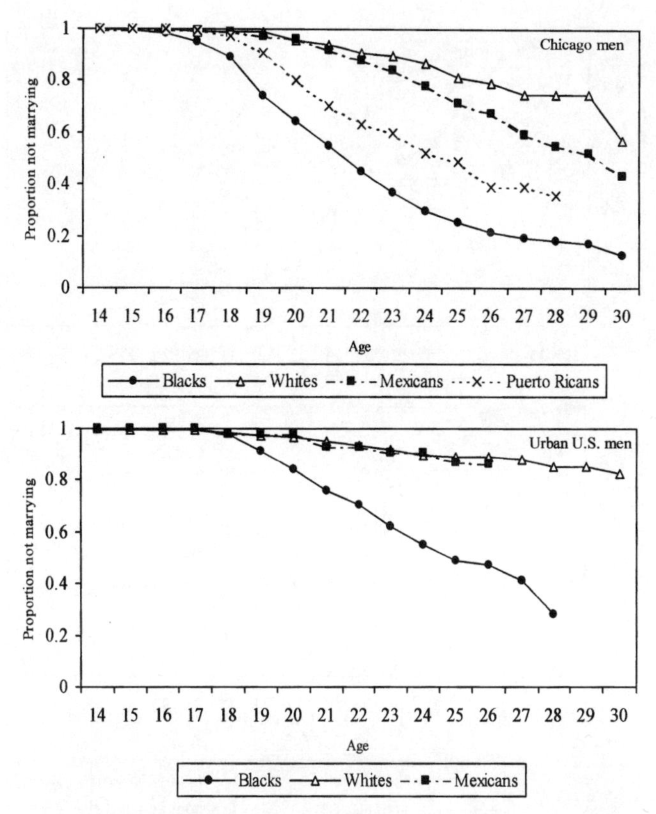

Figure 4.7 Birth Pathways to Family Life: Chicago and U.S. Urban Men by Race and Ethnicity

black men are substantially more likely to begin family life by fathering a child compared to other groups.

Racial differences in both the timing of marriage and births, coupled with the differences between Chicago and urban blacks nationally, suggest that family formation patterns observed in the inner city capture both race-specific and poverty-specific effects. In other words, differences in family formation patterns by race cannot be reduced to differences in poverty status alone. To examine this possibility, we compared the pathways to family life between black and nonblack mothers in Chicago and for the low-income population of U.S. central cities. We restricted the analysis to women because the sample size of poor urban men in the national sample was too small for subgroup analyses. Figure 4.8, which displays the survivor functions, reveals the additive effect of race and poverty on

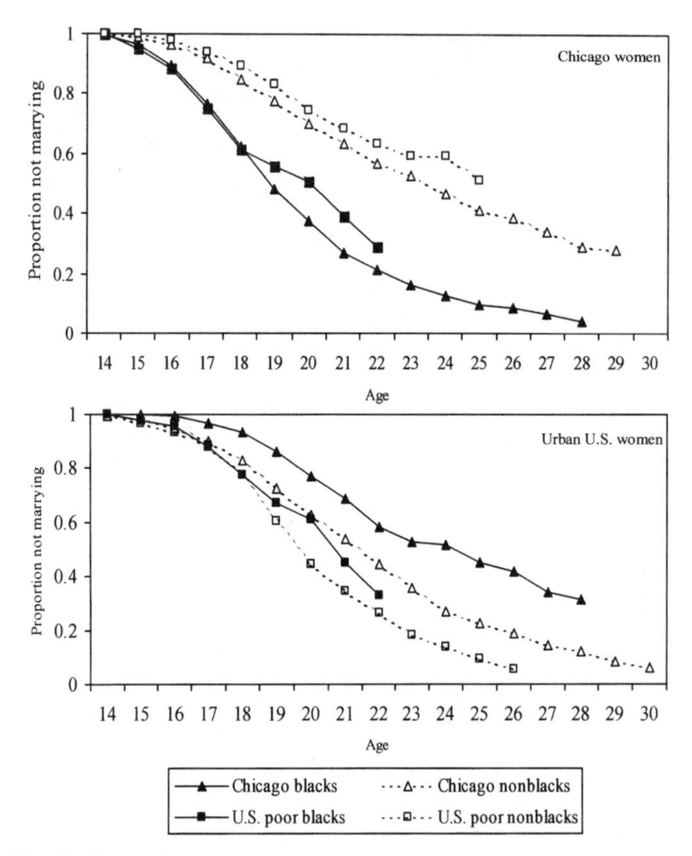

Figure 4.8 Marriage Pathway to Family Life: Chicago and Urban Poor U.S. Women

family behavior. Black mothers in Chicago exhibit somewhat lower rates of marriage than their poor, urban counterparts nationally, but they exhibit a pattern of family formation via premarital births that is strikingly similar to that of poor mothers residing in U.S. central cities. This pattern differs from that observed for poor nonblack mothers in both samples. These results lead us to conclude that inner-city mothers are not unique when compared to poor urban mothers, and that *both race and poverty contribute independently to the likelihood of entering adult family life via nonmarital fertility.*

To summarize, family formation patterns among Chicago's inner-city mothers and fathers differ from those observed among all urban parents. They confirm that nonmarital fertility is more prevalent in Chicago, and that it occurs earlier in the life cycle. Moreover, the proportions of parents who never marry are much higher among Chicago inner-city residents than among urban residents in general, especially for blacks. In these respects, our results are consistent with prior studies of poverty and family formation. However, our analyses go beyond existing studies in clarifying the intersection of race and class by showing that the experiences of Chicago's mothers are less atypical when compared to the low-income populations generally. Among poor people, marriage is no more common than an extramarital birth. The observed differences may also reflect the influence of family background and unequal opportunities afforded to residents of poor, inner-city neighborhoods. To scrutinize further our claims about independent racial and poverty effects on family formation, we turn to a multivariate analysis, which permits a more rigorous assessment of additional influences on pathways to family life. These analyses reveal whether the racial effect derives largely from the higher economic disadvantages experienced by inner-city black mothers, or whether race-specific effects persist across settings.

In examining the correlates of alternative pathways to family life, we emphasize the unique effects attributable to group membership and early experiences with poverty and parent absence. Because we focus on early disadvantages in life transitions, we examine the risk of entering family life via marriage or via a birth while the respondents were in their teens. To do so, we estimated a multinomial logistic regression analysis that predicts the odds of entering family life through either pathway at each age between thirteen and nineteen. The procedure is discussed in detail in appendix B, and the results of the statistical estimation are presented in appendix tables B4.4 (for women) and B4.5 (for men). The unique effects of each covariate on the probability of entering family life derived from these models used population characteristics reported in appendix table B4.6 and are presented in figures 4.9 to 4.11 below.

In general, the probability of a transition to family life during adolescence is relatively low, especially for men. Nonetheless, our findings reveal racial and ethnic as well as family background effects on pathways to family life. Racial and ethnic differences in the mode of entry to family life remain virtually unchanged after controlling for family background and early life experiences. As revealed in figure 4.9, blacks are significantly more likely than Hispanics and whites to enter family life via birth. Specifically, the average black woman is about 3.5 times more likely than her white counterpart to begin her family via childbirth. Moreover, black women are about 3.5 times more likely to enter family life via birth than via marriage. Premarital birth rates for black men in Chicago are seven times higher than for whites, and they are four times more likely to form

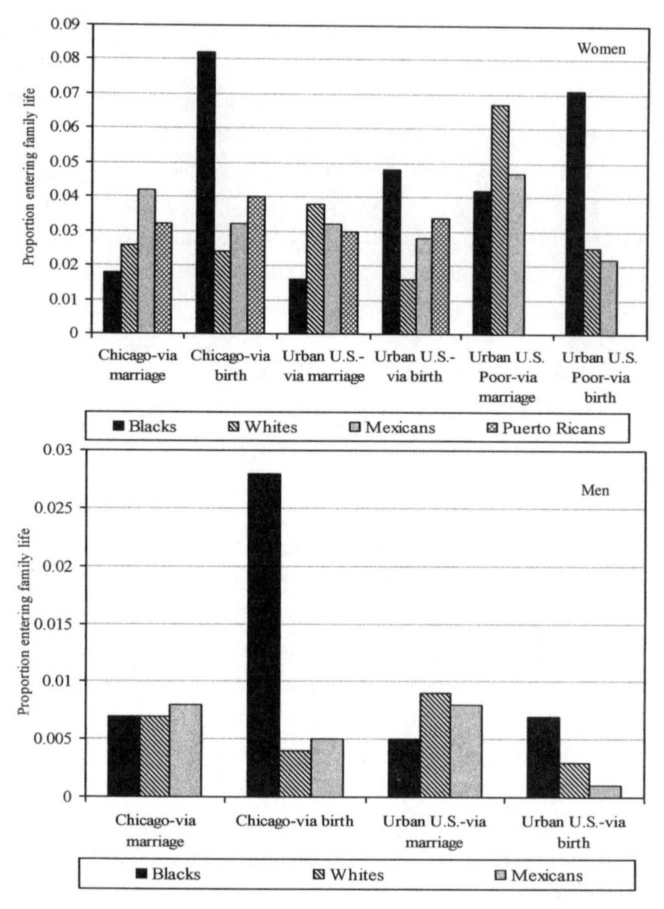

Figure 4.9 Racial and Ethnic Differences in Pathways to Family Life by Age 19: Chicago and Urban U.S. Women and Men

a family by fathering a child than by marrying before fathering children. Urban U.S. black men also have higher rates of premarital births compared to whites and Mexicans, but they are only 25 percent as likely as blacks in Chicago to experience this pathway to family life. These findings reinforce our claim that nonmarital births are the dominant pathways to adult family life for inner-city black women. Differences between Hispanic and white women are generally trivial except for higher premarital birth rates among Puerto Ricans. White and Puerto Rican women in Chicago are as likely to enter family via birth as via marriage, while Mexicans have the highest rate of family formation via marriage.

Estimates from the national survey reinforce our prior conclusions about the salience of race, but not Hispanic origin, in shaping family formation patterns. Especially noteworthy is that the pattern of racial effects on pathways to family life, although smaller, is identical to that observed in Chicago's inner city. Specifically, urban black women are three times more likely than white women with similar characteristics to enter adult family life via birth. Black women are also three times more likely to form families via a birth compared to marriage. Similar results obtain among the poor segment of the urban population, where the rates of premarital births for black women are only slightly lower than the rates observed for Chicago women. Compared to blacks, whites and Mexicans in the urban population, and more so among the poor, have a substantially higher tendency to enter family life via marriage than via birth. This indicates that poor urban whites and Mexicans experience poverty differently from blacks. In particular, the selection mechanisms into poverty may be less chronic and less associated with "concentration effects" than is true for blacks. This interpretation squares with evidence provided in chapter 2 from Chicago showing that inner-city neighborhoods inhabited by blacks are poorer and more highly segregated than those where whites and Hispanics live. Ghetto-specific behavior, to the extent that it exists, is the result of these two circumstances that are unique to blacks (Massey and Denton 1993).

Results for men also confirm that both race and poverty decisively influence family formation patterns. While black men have higher rates of entering family life via birth compared to all other groups, this difference is much more salient in Chicago than among urban men nationally, including poor urban men. Nonetheless, the probability that men of all racial and ethnic groups will enter family life as teens is very small when compared to rates for women. Because family background and early life experiences with poverty do not eliminate racial differences in pathways to family life, we conclude that *there is a race-specific component to family formation behavior.*

A second important finding is that disadvantages of the family of orientation reverberate for the family of procreation. As many studies have shown, growing up with a single parent influences the odds of beginning family life with a premarital birth, independent of minority-group status. As figure 4.10 shows, for both Chicago and urban U.S. women reared in single-parent families were 1.5 times (in the inner-city and among the urban poor) to 2 times (among urban women nationally) more likely to enter family life via births than women who grew up with two parents. A similar pattern is observed for urban men nationally, but not for

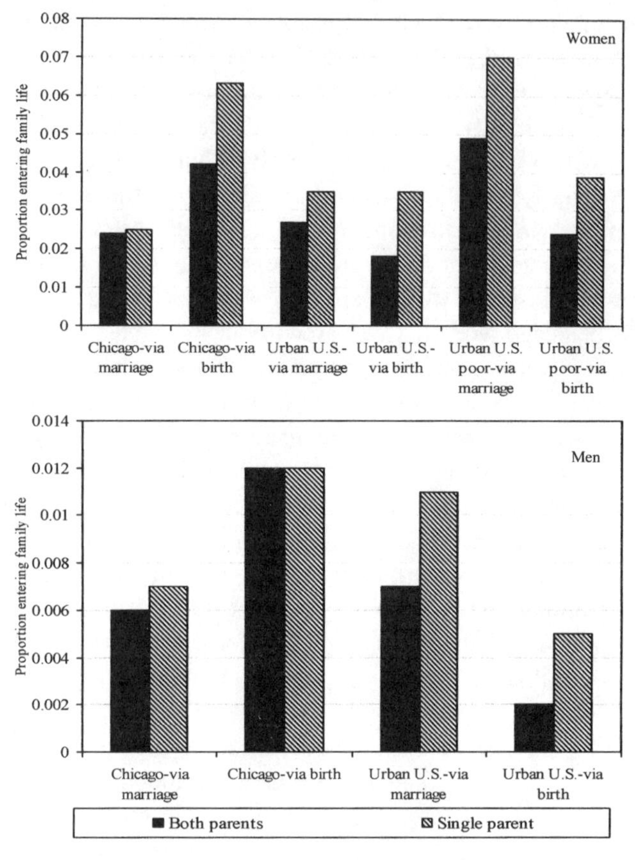

Figure 4.10 Family Structure Differences in Pathways to Family Life by Age 19: Chicago and Urban U.S. Women and Men

inner-city men. The former are more likely to enter family life before age twenty (either via birth or marriage) if reared by a lone parent, while for the latter there were no family structure differences in the probability of entering family life as a teenager. Among urban poor women, those reared in one-parent families were almost twice as likely as women raised by two parents to form families as teens. These findings further reinforce our claim that early entry into family life, either via marriage or birth, is associated with the correlates of poverty—in this instance, family structure.

Exposure to material deprivation during childhood also significantly increases the risk of entering family life through a birth, and this effect is particularly pronounced for Chicago's inner-city residents and the urban poor. As figure 4.11 shows, Chicago mothers reared in poverty are 1.7 times more likely to enter family life via an adolescent birth than those reared in a nonpoor family. Moreover, Chicago mothers are also less inclined to enter family life via marriage at any age. Although urban women nationally are substantially less likely to have a premarital birth compared to Chicago inner-city mothers, those exposed to welfare during childhood and adolescence were twice as likely as their counterparts whose families were never on welfare to experience a premarital birth. Growing up in economic deprivation also increases the probability that poor urban women will become (unmarried) teenage mothers, but early welfare exposure also increases their probability of marrying during adolescence.

To summarize, our multivariate analysis confirms what we have shown

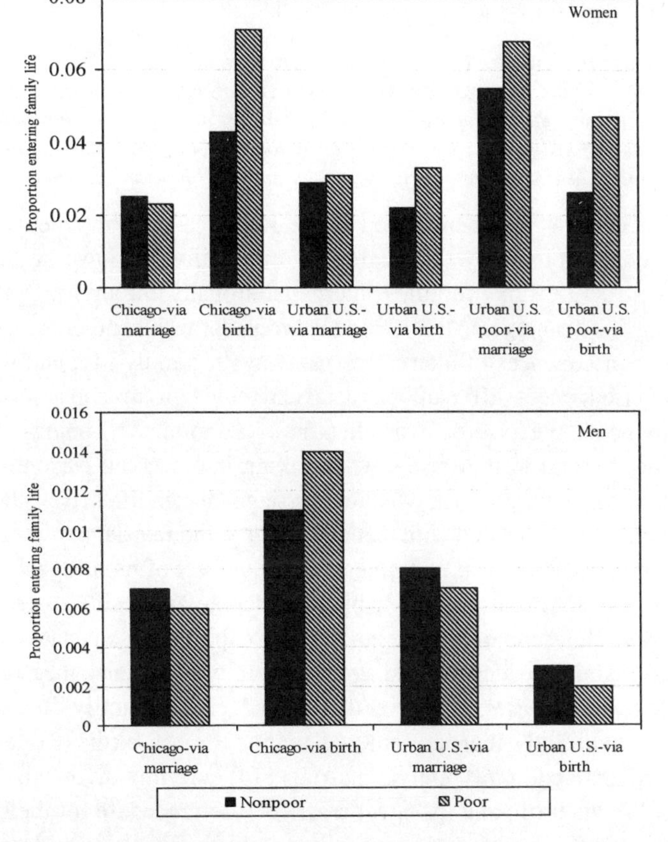

Figure 4.11 Family Poverty Status Differences in Pathways to Family Life by Age 19: Chicago and Urban U.S. Women and Men

and argued earlier: race and poverty are two important dimensions associated with the transmission of economic disadvantage from the family of orientation to the family of procreation. Both circumstances influence the likelihood of premarital births for women and men, but sex differences are highly significant and imply higher costs of the transition to family life for women than for men. Both marriage and the occurrence of a premarital birth during adolescence are associated with lower educational attainment and labor force participation. Although there is some debate about the causal relationships between educational attainment and adolescent childbearing and between women's assumption of family roles and their labor force activity (see discussion in Ahituv, Tienda, and Tsay 1998), there is no doubt that these behaviors place women at a higher risk of poverty in their families of procreation.

Consequences of Family Trajectories

> *The family structure is being broken down. I think it's going to have to change, 'cause otherwise our kids, they'll go through the same cycle that their parents did: dropping out of school, drugs, pregnancy, welfare, not seriously looking for a job, nor seriously looking out for the future, not setting goals. At the rate things going on now, it'll have to get worse.*

Billy, a chronically unemployed black father, asserts intergenerational consequences of family disintegration. Like Billy, we have argued that family structure links families intergenerationally through poverty and material disadvantage. In this section we consider whether there are, in fact, lasting consequences of alternative pathways to family life, and whether these consequences differ among racial and ethnic groups and across urban environments. We focus on two indicators of economic well-being—household income and welfare use—while taking into account current marital status. In particular, we ask whether marriage later in life offsets the early disadvantages associated with teenage fertility and marriage.

The empirical analysis examines the joint effect of pathways to family life and current marital status, which we denote as "family experience," on current household income and welfare participation status among groups of similar education and age. Our family experience measure consists of five categories: (1) currently married, entered family life via marriage; (2) currently married, entered family life via birth; (3) currently divorced, entered family life via marriage; (4) currently divorced, entered family life via birth; and (5) never married. The procedure for the statistical analysis is discussed in appendix B, and the results are reported in figure 4.12.

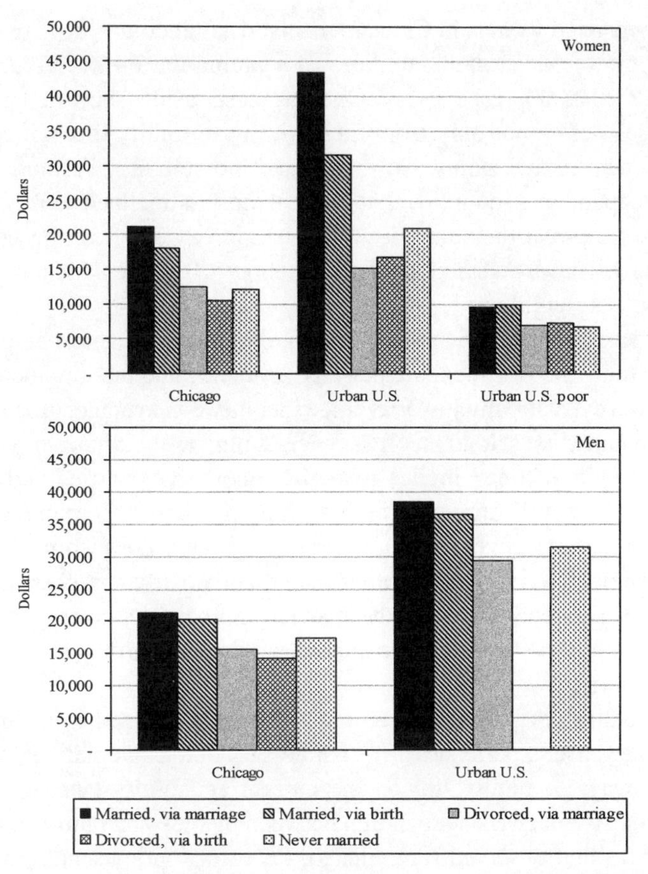

Figure 4.12 Adjusted Household Income by Family Experience: Chicago and Urban U.S. Women and Men

The left cluster of each part of the figure presents the adjusted means for household income of Chicago inner-city parents at the time of the survey. Married parents (men and women alike) enjoy the highest household income due both to men's higher earning capacity and the possibility of two incomes. Nonetheless, some income differences exist among married-couple households, depending on their early family experience. Chicago women who entered family life via marriage have a higher household income than their counterparts who entered family life via a birth ($21,000 versus $18,000, respectively). Similar differences obtain among divorced women; those who began families via marriage enjoyed higher household income than those who entered family life via birth, although both have substantially lower income than their married counterparts.

Never-married women in Chicago received a higher income than divorced women who began their families via a premarital birth—$12,000 compared to $10,000, respectively. Overall, these results show that economic penalties derive not only from the pathway to family life, but also from changes in marital status. Never-married mothers may be more resilient and independent—qualities that make them less inclined to rely on marriage to improve their families' well-being. Nonetheless, the household income of never-married mothers is about 70 percent that of married women with children.

Not surprisingly, the average household income of all urban parents is higher than that of Chicago inner-city residents, and the low-income stratum has a substantially lower average income.[4] On average, urban married women enjoy household incomes twice as high as their inner-city counterparts, and five times higher than the income of married, urban poor women. Across the three samples, striking parallels appear in the income of divorced and never-married women, although the income difference between divorced Chicago mothers and divorced urban mothers is smaller than the difference between the married mothers. On average, divorced and never-married women live in households with half the income of currently married women.

Especially noteworthy, given our substantive interest in documenting how early life experiences reverberate on adult outcomes, is the effect of pathways to family life on the current economic standing of urban mothers. Although the association between income and family experience resembles that observed for Chicago, the effect on household income of early family formation is more pronounced for the urban U.S. population. Specifically, household income of women who entered family via birth and subsequently married is 72 percent of the income of women who entered family life via marriage. The difference in the effect of family pathways between Chicago and all urban women derives from the higher level and greater income dispersion among the urban population. For the poor segment of the urban population, the only meaningful difference in household income is between currently married and unmarried women, independent of the pathway to family life.

Early family experience matters less for men's economic well-being

4. The lower income of the poor U.S. population is not surprising, given that the group was selected on the basis of income, while Chicago inner-city residents were selected according to their place of residence. Thus the income range and the variation are considerably higher in Chicago.

than for women's. Married men in both samples enjoy higher incomes than their unmarried counterparts, but the differences are smaller than those observed for women. Notably, pathway into family life is not significantly related to men's household income. These gender differences are not surprising given that women, as main providers for children, are more likely to bear the costs associated with divorce or premarital childbearing. Simply put, having a premarital birth constrains women's opportunities in the marriage market and the job market more than is true for men. Thus, in Chicago and among urban U.S. urban mothers nationally, being married, especially if family life was entered via marriage, provides the best shelter from economic hardships for the family of procreation.

Household income is only one measure of economic well-being. Therefore, and in keeping with our interest in intergenerational transmission of poverty, we examine whether early family experience has prolonged consequences for welfare participation, independent of respondents' race and ethnicity, age, and education (as detailed in appendix B). Adjusted welfare participation rates for women, reported in figure 4.13, show that divorced and never-married mothers are more likely to be on welfare than married mothers. However, within each marital status group, those who entered family via birth have higher rates of welfare use compared to otherwise similar women who began their families via marriage. These rates are particularly high among never-married mothers for all three subsamples. For example, among Chicago women, 62 percent of never-married mothers were on a welfare grant at the time of the survey, compared to 56 percent of divorced women who began their families via a birth, and 48 percent of divorced mothers who began their families via marriage. The corresponding figures for urban mothers are 45 percent, 33 percent, and 26 percent, respectively. As expected, currently married mothers have the lowest rate of welfare use, mainly through the AFDC-U (unemployed fathers) program. AFDC-U welfare episodes tend to be relatively short in duration. However, even among this group the rate of welfare receipt is lower among those who entered family life via marriage rather than via birth (18 versus 22 percent in Chicago, 6 versus 9 percent in the urban U.S., and 16 versus 24 percent among the urban poor). Place differentials in welfare participation are consistent with income differences described above, although it is noteworthy that these rates are very similar among Chicago and poor urban mothers.

For men the relationship between family experience and welfare dependence is different. Divorced men in Chicago, especially those who entered family life via birth, have the highest rates of welfare reliance (27 and 34 percent via marriage and birth, respectively). No differences in welfare participation rates by family pathways are observed among married men in Chicago. The rates of welfare participation are very low among urban U.S. men, and the differences by marital status or by pathways to family life are tiny. However, men who entered family via birth (whether currently married or never-married) have higher rates of welfare participation compared to those who entered family in the traditional way.

Last, we examine racial differences in the effect of family experience on household income and welfare participation. Having established that blacks experience a higher risk of entering family life via births, we consider whether the lower economic standing (both in income and in welfare

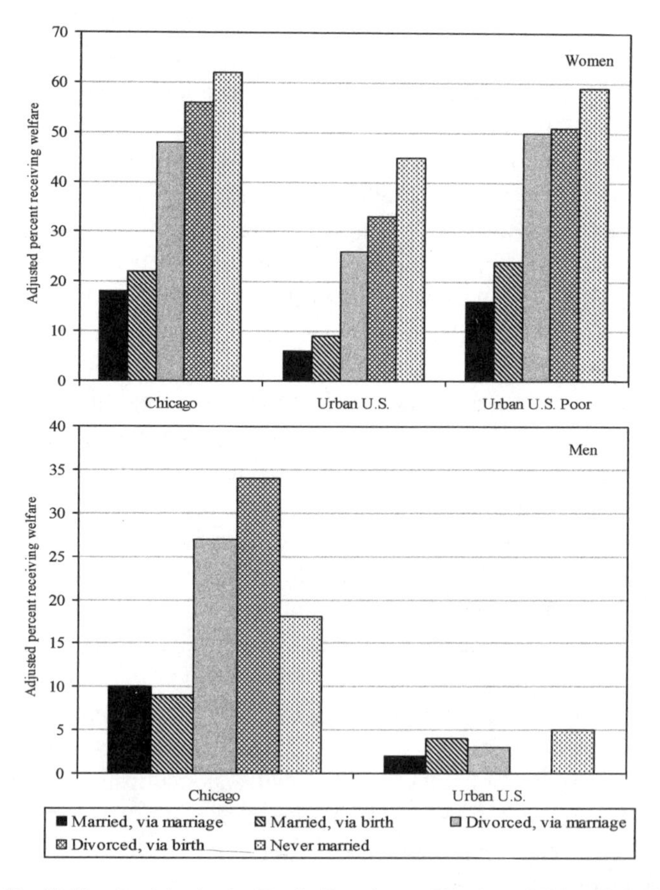

Figure 4.13 Welfare Participation by Family Experience: Chicago and Urban U.S. Women and Men

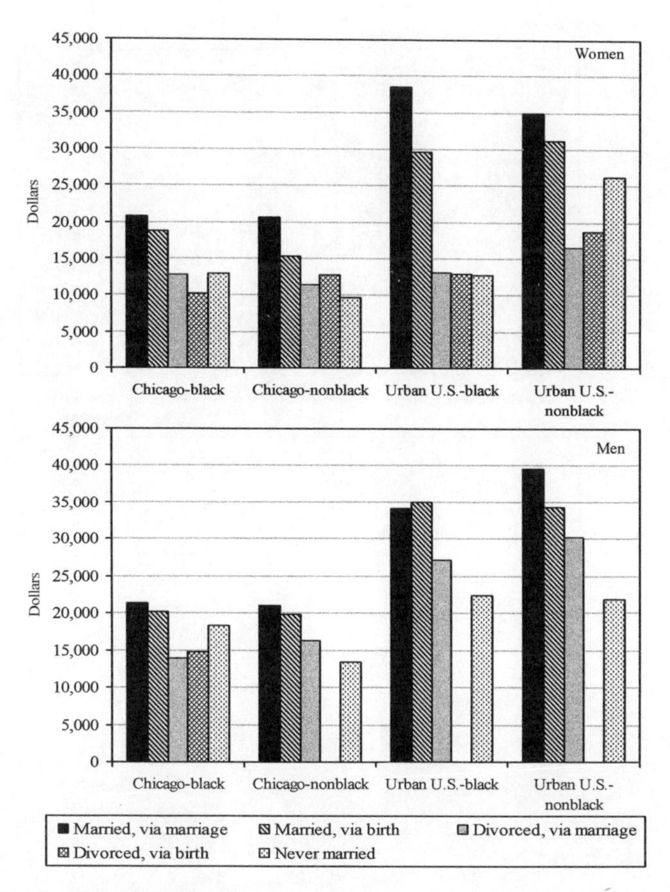

Figure 4.14 Adjusted Household Income by Family Experience and Race: Chicago and Urban U.S. Women and Men

participation) of those entering family via birth is due to their early family experiences (i.e., having a premarital birth) or to their race. Figure 4.14 presents the adjusted means of household income for black and nonblack women and men. The figure shows that the general pattern of income differences by marital status and family pathways are similar across settings for black and nonblack parents, albeit somewhat less consistent for men.[5] Most striking is the *absence* of racial differences in household

5. Among married, urban, black women, the high income level and the relatively large difference in household income between those entering family via marriage and those entering via birth is probably due to the small number of women in these categories. It is therefore hard to draw firm conclusions from this finding.

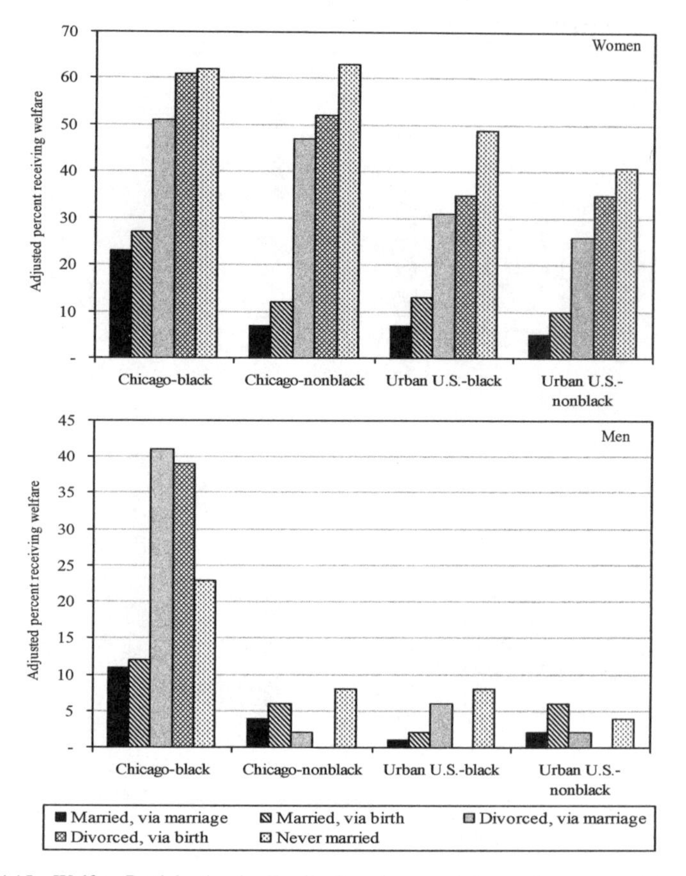

Figure 4.15 Welfare Participation by Family Experience and Race: Chicago and Urban U.S. Women and Men

income across settings. Thus, we infer that *family experience* and the *context* of poverty in which parents forge their livelihood rather than race per se are responsible for the observed differences in household income.

The pattern of association between family experience and welfare dependence across racial lines is somewhat different from that of income (see figure 4.15). The general effect of family experience remains unchanged for all groups. That is, married women are less likely to use welfare than divorced or never-married women. Also, within each marital status category, women who entered family via birth exhibit higher welfare use rates than those who entered via marriage. These findings obtain for men as well. However, the major difference from the income pattern is the emergence of racial differences in welfare use among Chicago parents.

Married black women who entered family life via marriage are 3 times more likely than their nonblack counterparts to rely on welfare for income maintenance; similarly, married urban black women nationally who entered family via birth are 2.3 times as likely as their nonmarried counterparts to receive welfare. Also, welfare participation rates among divorced black women are higher than among their nonblack urban counterparts, independent of their pathway to family life. These findings lend support to W. J. Wilson and Neckerman's (1986) male marriageable pool hypothesis inasmuch as black inner-city women gain lower benefits from marriage compared to nonblack women. Given the high concentration of blacks in ghetto neighborhoods, and the limited availability of "marriageable men" (adequately employed), these findings may also suggest that residents of ghetto-poverty neighborhoods develop a "a taste for welfare." We analyze the pattern of welfare use, including preferences for welfare, in the next chapter.

Summarizing, we examined the long-term effects of pathways to family life on the economic standing of women and men in our three subsamples, and empirical results suggest that having a premarital birth has lasting economic consequences—more so for women than men. These effects persist even after women marry, and are fairly consistent across settings and according to group membership. Furthermore, the findings reported in this chapter suggest that in the context of poor neighborhoods, significant racial differences exist in welfare participation, which we examine at greater length in chapter 5.

Conclusions

> *I think that society should reexamine its values. Uh, a lot of things within society need to be changed and be redefined. Like the family. . . . I was raised in a one-parent family. Society should recognize that these families exist, one-parent family, and it should be different goals and values for those families, not like two-parent families. . . . I think that they need to start with the family, and that's the only way they're gonna change things. If somebody writes a book, finds a better way of raising a family. Teach the people, teach them a better way.*

Shala, thirty-seven, a never-married, black mother of two, has a college degree, but she is chronically unemployed. She perceives the family as one of the most important social institutions that influence the life chances of youth, and she worries about what she perceives to be a breakdown of the black family. She uses her own experience as a point of reference between then and now.

> *Where I was growing up, where I was living in public housing, things weren't right. I looked around and I said, "What the fuck is going on?" Families don't stay together, people were having babies like popcorn, kids weren't going to school. I didn't like it. That's the life I saw. . . . It has a lot to do with society's attitudes toward black family, the black family's attitude about family life itself: they don't seem to know as much as they should know. It seems that it's children raising children.*

In this chapter we examined the early beginnings of Chicago inner-city parents, focusing on educational attainment and family formation, and their consequences for adult well-being. Our findings confirm that for youth reared in poverty and with lone parents, disadvantage is accumulated through the life course and transmitted intergenerationally. Adolescent experiences have prolonged consequences for adult lives. We show that both race and poverty are decisive in shaping adolescents' problem behavior and their pathways to family roles, and we document significant racial and ethnic differences in family background. We also demonstrate its importance to young adults' behavior. In particular, we find that, on average, blacks and Puerto Ricans begin life in more disadvantaged circumstances than whites or Mexicans, and that racial and ethnic differences persist across samples. We argue that racial and ethnic inequality during adolescence derives not only from group membership per se, but also from the circumstances that delimit opportunity during adolescence and beyond. Disadvantaged families of orientation determine two pivotal life events for adolescents, namely, high school graduation and the pathways into family life, and these life choices have lasting economic consequences through adulthood.

Racial differences in pathways to family life are especially noteworthy. We show stronger racial effects on pathways to adult family life for the Chicago sample than for the national urban sample. The age pattern of racial differentials reflects the extremely high incidence of teen parenting in inner-city ghettos (Hogan and Kitagawa 1985; Anderson 1991, 1990). However, this finding does not necessarily indicate the existence of ghetto-specific behavior, because blacks are disproportionately concentrated in high-poverty neighborhoods (Massey and Denton 1993). Therefore, the high rates of teen pregnancy in urban ghettos may reflect the sorting of poor blacks into concentrated poverty neighborhoods, which in turn perpetuate syndromes of disadvantage among youth reared in extreme material disadvantage (Tienda 1991).

In addition to racial and ethnic variation in poverty-related behaviors, we find large sex differences in pathways to family life, and we document their prolonged implications for the economic well-being of women and children. Among all racial and ethnic groups, women are at a higher risk of experiencing premarital parenthood early in life than men. Moreover, premarital parenthood has lasting economic consequences, especially for women, regardless of whether they eventually marry, or divorce subsequent to a birth. Thus, Hogan and Astone's (1986) conception of a "pathway" as a sequencing of life-course events with implications for future transitions and outcomes has strong empirical foundation in Chicago's inner city.

CHAPTER FIVE

Doles and Safety Nets: Public Assistance and Income Support

I'm unemployed, I have two children, I'm on public aid, and I am willing to work. And like I said . . . it's . . . it's all in the system, because I feel like . . . I have to take care of two children, I'm a single parent, and no job is gonna get me no . . . I could take a job, I can work at McDonald's, but that's not gonna have me take care of my children, that's not gonna buy clothes and food, public aid doesn't . . . and medicine. Public aid doesn't buy it either, you know, but . . .

Cassandra, twenty-eight, a black mother of two who never married or completed high school, portrays the harsh realities of being poor. Having experienced chronic unemployment and welfare dependence, she understands that minimum-wage jobs seldom provide a better alternative to welfare, especially if they offer no benefits. And as a single mother struggling to make ends meet, she knows firsthand the difficulties of raising children on meager welfare checks. For thousands of women like Cassandra, AFDC is the only viable way to make a living. Mary, thirty-four, a never-married, black mother of three who has been chronically dependent on AFDC, agrees. For her, public aid is opportunity, but only to subsist: "People in this neighborhood, well, well . . . they have opportunity through aid. Without aid they would fall, without aid we would all fall."

Both Cassandra and Mary live in Chicago's poorest neighborhoods, where nonmarital childbearing and reliance on public income support are pervasive, and working adults are the exception rather than the norm. The presumption, of course, is that Cassandra and Mary do not want to work—that their dependence on AFDC is voluntary and that they prefer welfare over work. Presumably, such attitudes and behavior distinguish the urban underclass from the poor in general. Yet, despite numerous attributions about welfare participation among the inner-city poor, systematic comparisons between the so-called urban underclass and the urban poor in general are lacking.

Accordingly, in this chapter we describe and analyze the pervasiveness of welfare utilization among inner-city residents, considering differences among black, white, Mexican, and Puerto Rican men and women. Comparisons with the national urban sample and the poor urban sample permit us to address whether and in what ways the welfare behavior of the inner-city poor is unique. Continuing with the theme of accumulating disadvantages over the early life course developed in chapter 4, we examine whether, to what extent, and in what ways early life events and disadvantages carry over into adult life and reproduce poverty and welfare reliance intergenerationally. This highly controversial question resides at the core of the debate about whether chronic welfare utilization results because poor mothers develop a "taste" for welfare, or whether prolonged dependence merely reflects the intergenerational transmission of material deprivation.

The chapter is divided into three sections. The first focuses on current and recent welfare participation for both the Chicago and national urban subsamples. Section 2 addresses the most politically sensitive issue, namely, the chronicity of welfare use. These analyses focus on conditional probabilities of leaving a welfare spell and the duration of welfare spells. The final section draws on the rich attitude and perception data available in the UPFLS and examines group variation in preferences for welfare. To conclude, we recapitulate key findings and pose questions about whether welfare recipients really want to work, which leads us into chapter 6.

Not surprisingly, we find higher rates of current and recent welfare use in Chicago compared to all urban areas, but Chicago's rates of recent welfare use were similar to those of the urban poor nationally. There are more similarities than differences in the determinants of welfare use across groups and settings, but large differences between men and women. Also, nationally as well as in Chicago, blacks and Puerto Ricans had higher rates of welfare participation compared to Mexicans and whites. However, racial and ethnic differences in recent welfare use largely reflect group variation in characteristics and early life circumstances that qualify them for public assistance, notably, having been reared in a mother-only family, childhood material deprivation, and the two pivotal life-course events that shape adult well-being: low educational attainment and bearing children out of wedlock. The noteworthy exception is that Chicago's black and Puerto Rican parents are more likely to receive welfare than similarly endowed whites or Mexicans. We attribute these differences to the sorting processes that historically confined blacks and to a lesser extent Puerto Ricans to the poorest neighborhoods rather than to group-specific preferences for welfare.

Analyses of the chronicity of welfare use also reveal racial and ethnic differences in patterns of welfare utilization. Again, black and Puerto Rican women experienced the longest spells and the slowest pace of exiting welfare, conditional on receiving a grant. However, even among poor, inner-city populations, over 40 percent of welfare episodes lasted two years or less. Moreover, the likelihood of a long duration was decisively influenced by whether women entered family life via marriage or births, reinforcing the detrimental effects of early disadvantage and the accumulation of hardships through the life course. Analyses of attitudes toward welfare reveal that the majority of respondents preferred a job to welfare, although responses vary according to current and past welfare status.

Public Assistance and Its Beneficiaries

"Welfare" typically refers to cash assistance provided by the federal government to needy individuals and families, but occasionally includes other types of income subsidy programs such as General Assistance (GA). The latter includes any cash assistance—usually temporary—to needy individuals and families who are ineligible for federal income support programs. However, in popular usage, and until 1996, "welfare" has largely referred to means-tested income provided to single mothers with dependent children. Aid to Families with Dependent Children, or AFDC, was established as part of the Social Security Act of 1935 to assist widows left alone to raise children. Conceived as a federally provided social insurance program to protect children against extreme poverty, AFDC (initially called ADC) allowed poor, single mothers to remain at home and raise their children until the age of eighteen. Improvements in health and safety coupled with higher rates of divorce and nonmarital childbearing shifted the composition of program beneficiaries from widows, the initial intended beneficiaries, to divorced and never-married mothers. In the minds of legislators and the general public, this demographic change in caseloads represented a shift from deserving to undeserving poor. Consequently, the AFDC program became increasingly controversial.

As welfare rolls swelled, AFDC benefits became steadily less generous, while eligibility requirements became more stringent. Since the 1960s, Congress has been concerned that the welfare system fosters dependence by discouraging work effort, by encouraging out-of-wedlock parenting, and by disrupting families (Duncan, Hill, and Hoffman 1988). These arguments were forcefully articulated in 1984 by Charles Murray, who blamed rising levels of black male joblessness, high black teenage birth rates, and rising female headship rates on public assistance programs. The more benign view of AFDC focused on the investment in children who, in any event, were not to blame for their parents' unwillingness or inability to hold down a job.

As controversy mounted over the putative negative consequences of AFDC on work effort and family disruption, the program became more bureaucratic and essentially trapped single mothers with children into poverty through work disincentives and extremely low benefits. For example, in 1987 single Illinois mothers with two children received a monthly AFDC check of $342, which is considerably more generous than the $118 paid to a similar family residing in Alabama, the lowest benefit state in that year, but well below the $749 available to identical families in Alaska, the most generous state in that year (U.S. House 1987, pp. 405–10, tables 8 and 9). Nationally, the average grant for a family of three in 1987 was a meager $354 per month, which was supplemented by an average food stamp benefit of $182 (U.S. House 1987, p. 407, table 8). The combined benefits represented just 47 percent of the 1986 poverty threshold, which virtually guarantees that AFDC families will remain impoverished.

Despite a voluminous research literature showing that chronic welfare dependence was the exception, not the rule, and that most spells of AFDC lasted under two years, concerns about work disincentives and family disruption effects of public assistance to single mothers persisted. The growing criticism of AFDC—both its bureaucratic aspects and failure to eliminate poverty—culminated in its abolition with the enactment of the Personal Responsibility and Work Opportunity Reconciliation Act of 1996 (PRWORA).[1] PRWORA effectively dismantled AFDC and imposed strict time limits on welfare receipt under the block-grant system that replaced it, Temporary Assistance for Needy Families (TANF). However, our analyses are based on the pre-reform period. Because much of the controversy about the emergence and growth of an urban underclass revolved around the twin pillars of chronic welfare dependence and joblessness, we consider whether and in what ways the welfare participation of the inner-city poor differs from that of the general urban population and the urban poor in particular. Specifically, we address whether the behavioral correlates of welfare participation differ among the inner-city poor

1. PRWORA abolished AFDC and replaced it with block-grant welfare funding for states, known as Temporary Assistance for Needy Families (TANF). While the target recipients for TANF—single mothers and their children—are the same as the beneficiaries of AFDC, the new block grants oblige states to move recipients into work activities, create incentives to reduce the welfare rolls, and impose time limits on welfare receipt.

and all poor urban parents, and we document the extent to which welfare dependence was chronic among AFDC recipients. By focusing on the most disadvantaged populations, our analyses not only address several issues left unresolved by the underclass debate, but also check the wisdom of the PRWORA reforms. The latter issues are discussed in the concluding section of the chapter and again in chapter 7.

Welfare Participation in Chicago and U.S. Central Cities

How pervasive is welfare participation in Chicago's inner city, and how different is it from the welfare behavior of urban parents nationally or the urban poor in particular? To address this question, table 5.1 compares three samples using two measures of welfare participation. One focuses on welfare use at the time of the survey; the other examines welfare use during the five years prior to the interview date. Welfare refers to episodes of general assistance and AFDC, with the latter almost exclusively used by women, and the former by men. The "recent use" measure gives a fuller picture of the pervasiveness of welfare utilization because most welfare episodes are short and thus not likely to be captured at an arbitrary survey date.

That inner-city parents exhibit higher rates of welfare participation than all urban parents is hardly surprising given the Chicago study design. In 1987, 16 percent of fathers received welfare payments compared to only

Table 5.1 Current and Recent Welfare Participation Rates: Parents Residing in Chicago's Inner City and Urban United States (Percent)

	Chicago		Urban United States		Urban U.S. Poor	
	Men	Women	Men	Women	Men	Women
% Current Users	16.0	45.0	2.3	12.9	9.8	39.7
Blacks	22.3	51.9	3.9	27.5	—	54.5
Whites	6.3	29.7	1.7	7.4	—	32.3
Mexicans	2.6	13.5	2.9	16.4	—	31.7
Puerto Ricans	11.7	53.7	—	34.0	—	—
% Recent Users	28.7	56.6	5.7	24.0	21.3	58.7
Blacks	37.5	62.2	7.5	38.0	—	65.9
Whites	19.6	33.9	5.0	17.8	—	57.1
Mexicans	5.7	21.2	5.9	33.2	—	50.8
Puerto Ricans	20.7	60.2	—	43.5	—	—
N	811	1523	947	1887	137	600

Source: UPFLS and NSFH

2.3 percent of urban fathers nationally. Women's welfare participation rates at the time of the survey were considerably higher. Almost half of all inner-city women, compared to 13 percent of urban women nationally, were on a welfare grant (mainly AFDC) in 1987. Welfare participation rates are higher based on recent use, that is, within five years of the survey date. Over one in four of Chicago's inner-city fathers reported having received General Assistance during the five years prior to the survey, compared to less than 6 percent of urban fathers nationally. Twice as many Chicago mothers as all urban mothers reported having received welfare during the five years prior to the survey date (57 and 24 percent, respectively). The small difference between recent and current use among Chicago mothers (45 versus 57 percent, respectively) suggests higher levels of chronic dependence in the inner city, which we demonstrate below.

A comparison between Chicago parents and the poor substratum of urban parents reveals striking similarities in recent welfare use. However, the cross-sectional measures (current use) show that welfare participation rates in Chicago's inner city were higher than among poor urban parents nationally. For example, less than 10 percent of urban poor fathers were on a welfare grant at the time of the NSFH interview, compared to 16 percent of Chicago fathers. Again, this probably reflects the higher levels of chronic use in the inner city.

Although the level of welfare participation of Chicago residents is generally high compared to residents of all central cities, there are marked racial and ethnic differences between the two populations. Blacks and Puerto Ricans exhibit the highest rates of recent welfare participation both in Chicago and in all U.S. central cities. Nearly one-quarter of black men in Chicago were on a welfare grant in 1987, and over one in three were recent welfare users. Only 12 percent of Chicago's Puerto Rican men, 6 percent of white men, and less than 3 percent of Mexican men were on a welfare grant when interviewed. Given low average rates of welfare participation among men nationally, racial and ethnic differences in welfare use are less striking than those among women, but nonetheless discernible. Nationally, urban black men exhibit higher rates of welfare utilization than whites or Hispanics. But in Chicago, black men were five times more likely to have received a welfare grant during the five years prior to the survey compared to their national racial counterparts who lived in central cities. This could reflect greater levels of need, that is, higher income shortfalls, or greater preferences for welfare—a question we pursue in the next section.

Racial and ethnic differences in women's welfare utilization are similar to those for men, inasmuch as black and Puerto Rican mothers exhibit the highest participation rates, both in Chicago and among all central-city dwellers. More than half of all black and Puerto Rican mothers in Chicago were on a welfare grant when interviewed in 1987 compared to 30 percent of their white counterparts and less than 14 percent of Mexicans. Among the urban sample, whites have the lowest rates of welfare participation, but in Chicago Mexicans are the least likely to use welfare. This anomaly reflects the immigrant composition of Chicago's inner-city, Mexican-origin population, the majority of which is foreign-born and largely married. Because eligibility for AFDC depends on marital status—specifically, being unmarried with dependent children—the marital status composition of Mexican immigrants means they are less likely to be eligible for these transfer benefits unless their spouses are unemployed. That Mexican husbands have high rates of labor force participation disqualifies this group for AFDC-U, a special program for families with unemployed fathers. A comparison to the national urban poor sample reinforces this interpretation. Among urban poor mothers, almost one-third of Mexican women received welfare at the time of the survey, and half of them had received a welfare grant during the previous five years. In fact, their rate of dependence is similar to that of urban poor white women. This is because immigrants represent a lower share of the urban poor, Mexican-origin population nationally.

Three sets of circumstances can account for differences in welfare participation among and across settings, namely, differences in characteristics that qualify residents for public assistance; differences in labor market opportunities that enable families to be economically self-sufficient; and differences in preferences for welfare over work. Although the public debate and media portrayals emphasized the last explanation, empirical researchers generally underscore group differences in socioeconomic composition and labor market opportunities, noting that welfare participation is an option only for the economically disadvantaged who meet strict eligibility guidelines. Differences in preferences are relevant mainly for families that qualify for aid. As primary caretakers, mothers are more likely than fathers to qualify for public assistance; hence, differences in eligibility for public aid largely explain the large sex differences in participation.

Evidence reported in chapters 3 and 4 showed that Chicago inner-city parents are from more disadvantaged family backgrounds than urban parents nationally, in that they are more likely to have been reared by single mothers and parents lacking a high school education. Similarly, higher shares of Chicago parents compared to all urban parents were exposed to welfare during their childhood. For example, 40 percent of mothers who had received public assistance during the five years prior to the survey reported having been exposed to welfare during at least half of their childhood, compared to approximately 20 percent of all urban mothers who were recent recipients. Also, welfare recipients were less likely to be married, and in particular, never married, than nonrecipients. This fact—the rise of beneficiaries who bear children out of wedlock—was one of the principal criticisms of the AFDC program that ultimately led to its demise.

W. J. Wilson (1987, 1996) made powerful arguments that concentrated urban poverty derives from the lack of job opportunities, and chapter 2 has documented the close correspondence between the changing spatial distribution of job growth and the rise of ghetto-poverty neighborhoods. A comparison of the welfare participation rates between the inner city of Chicago and the national urban poor population sheds light on the distinction between poor people and poor places. In view of allegations that residence in concentrated poverty neighborhoods perpetuates a culture of deviance that putatively encourages but also sustains chronic welfare use, it is relevant to consider how welfare participation varies according to neighborhood poverty. The findings presented in table 5.2 confirm that welfare use increases dramatically as the neighborhood poverty level rises for most (but not all) groups. In the poorest neighborhoods (40+ percent poverty), which are predominantly black neighborhoods, more than 70

Table 5.2 Welfare Participation Rates by Neighborhood Poverty Rate: Women and Men in Chicago's Inner City

	Neighborhood Poverty Rate			
	<20	20–29	30–39	40+
Women				
Blacks	12.0	38.6	57.2	72.0
Whites	16.9	36.4	33.3	—[a]
Mexicans	5.7	13.5	19.6	—[a]
Puerto Ricans	—[a]	53.3	52.3	—[a]
N	161	732	379	251
Men				
Blacks	8.7	19.0	26.2	29.6
Whites	3.8	2.1	—[a]	—[a]
Mexicans	2.4	2.3	4.5	—[a]
Puerto Ricans	—[a]	10.6	12.9	—[a]
N	130	402	186	93

[a] Too few cases.

percent of the black women were on a welfare grant when surveyed, but only 12 percent of black women who resided in neighborhoods with a poverty level below 20 percent received AFDC. This is because black women who live in low-poverty neighborhoods are themselves less likely to be poor and thus not to require income transfers.

Racial and ethnic differences in men's welfare use across the different neighborhoods reveal that blacks have the highest welfare utilization rates in all neighborhoods, followed by Puerto Ricans. Among women, however, racial and ethnic differences in welfare participation across neighborhoods are less consistent. In neighborhoods where 30–39 percent of residents are poor, more than half of the black and Puerto Rican women were on public assistance, compared to one-third of whites and only 20 percent of Mexicans. In places with lower poverty, however, blacks are as likely as (or even less likely than) whites to receive welfare, while Puerto Ricans have the highest rate of welfare use. For example, half of the Puerto Rican mothers residing in neighborhoods with 20–29 percent poverty compared to one-third of black and white mothers were on public assistance at the time of the survey.

The racial and ethnic differences in welfare participation across neighborhoods largely reflect the forces of segregation—the fact that whites are less likely than blacks and other minorities to live in the poorest neighborhoods—and the economic restructuring that transformed Chicago's inner city, as described in chapter 2. Because Chicago is the most segregated city along racial and class lines, the few blacks and Puerto Ricans who manage to locate in low-poverty neighborhoods may be better off financially, on average, than whites who live in poverty neighborhoods. Hence, their welfare participation rates are lower. The lower welfare rates may also capture sampling variability and the small sample sizes of blacks residing in white neighborhoods.

Racial and ethnic differences in welfare participation rates also could reflect group differences in *preferences* for welfare or group differences in levels of *disadvantage* that eventuate in high rates of welfare participation. Do the observed differences reflect group-specific or ghetto-specific behavior? Because blacks, whites, Mexicans, and Puerto Ricans differ in several characteristics that influence eligibility for welfare benefits (and where they can afford to live), including educational attainment, family background, early exposure to poverty, and the prevalence of nonmarital childbearing, it is not possible to infer the existence of group-specific or ghetto-specific welfare behavior based on aggregate rates. In most of these characteristics that transmit poverty intergenerationally, blacks experience the greatest disadvantages. Therefore, it is important to separate the two effects—group membership and disadvantaged background—to ascertain whether racial and ethnic differences in welfare participation persist after taking into account the share due to group differences in early and accumulated disadvantages.

Multivariate Analyses of Recent Welfare Use

To address whether racial and ethnic variation in recent welfare participation are attributable to differences in early and accumulated disadvantages, we used discrete event-history methods and considered all welfare episodes over a five-year period. Appendix B reports the procedure of file construction and model-specific details. The results of the logistic regression analysis are presented in tables B5.2 (for women) and B5.3 (for men). All coefficients were converted to probabilities using the sample means reported in table B5.4 for ease of interpretation and are depicted graphically in the following sections. Thus, the bar graphs in figure 5.1 represent adjusted probabilities of welfare use by race and ethnicity.

Figure 5.1 reveals that racial and ethnic differences in recent welfare participation rates among all urban women virtually *disappeared* after controlling for family background and early life events that determine eligibility for and propensity to use public assistance. Stated differently, for all central cities, we find *no evidence of group-specific welfare behavior:* all groups were equally likely to have received public assistance during the five years prior to the survey if they had similar background characteristics and comparable levels of early material disadvantage. Among the national urban samples, the main differences between all central-city residents and poor residents is the *level* of welfare use, which is striking, even if consistent with expectations. The urban poor are *three times* more likely to receive public assistance than all central-city residents. Whereas welfare participation rates of all urban women were consistently under 10 percent for all groups considered, they hovered around 28 percent for poor urban women. In Chicago's inner city, only white women's welfare participation rate was comparable to that of urban poor women nationally, and the participation rate of Chicago's Mexican-origin women was similar to that of all Mexican women residing in central cities. Conceivably, the low participation rate of Mexican-origin women in Chicago reflects their immigrant composition, but for urban Mexican-origin women nationally, income and marital status figure more prominently in keeping welfare rates low.

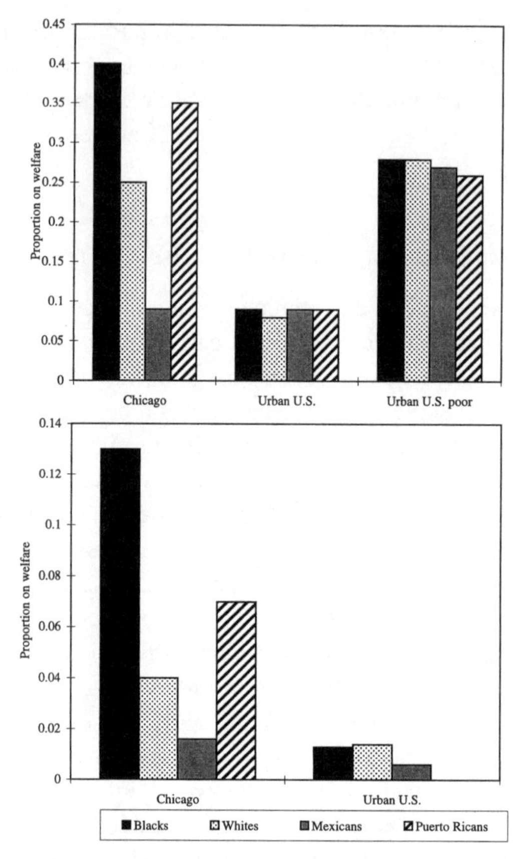

Figure 5.1 Risks of Recent Welfare Use by Race and Ethnicity: Chicago and Urban U.S. Women and Men

For Chicago women, however, racial and ethnic differences in welfare participation still *persist* even after taking into account relevant background characteristics and early experiences that influence eligibility for and receipt of public assistance. That is, Chicago black and Puerto Rican women who resided in poor, inner-city neighborhoods *are* more likely to use welfare than white women with similar characteristics who also live in poor neighborhoods, and Mexican-origin women—mainly immigrants—are less likely to receive income transfers than comparably situated white women. We could not control for neighborhood poverty in the multivariate model because this attribute was measured only at the time of the survey, and our measure of welfare use refers to the five years prior to the survey. Field procedures revealed that the inner-city population experiences a considerable amount of residential mobility—more than nonpoor residents. To properly consider neighborhood effects requires time-varying information about neighborhood of residence throughout the five years, which this survey unfortunately lacks. Men's probability of recent welfare use is considerably lower than women's across settings, as shown in figure 5.1. However, as observed for women, among the national urban population, racial and ethnic differences in men's recent welfare use are trivial. This means that the observed differences in table 5.1 arise from group differences in the likelihood of being poor rather than group-specific differences in tastes for and propensity to use welfare. However, among Chicago's inner-city fathers, blacks are four times more likely than whites to have received welfare during the five years prior to the survey. These differences in welfare participation in Chicago defy easy explanation and have fueled arguments about the emergence of a welfare culture that values public assistance more than work.

In interpreting these results, we should emphasize that poor white and Hispanic neighborhoods are nowhere near as poor as impoverished black neighborhoods, and that many blacks reside in concentrated poverty neighborhoods because they live in subsidized public housing. This leaves open the possibility that these differences reflect ghetto-specific behavior, and that the welfare behavior of the urban underclass is distinct from the that of the urban poor in general.

The differences among the three samples underscore the importance of place in organizing, if not shaping behavior, because, as shown in chapters 2 and 3, blacks in Chicago are considerably more likely to live in extreme poverty neighborhoods. And only in the extreme poverty neighborhoods do we observe extremely high levels of welfare participation (table 5.2). Yet interpreting these differences as reflecting ghetto-specific behavior is not entirely satisfactory, because the sorting processes that confine blacks to the poorest neighborhoods have more to do with housing policy, historic and contemporary segregation, institutional discrimination, and accumulated disadvantages than with preferences for welfare or ghetto-specific behavior. We return to the issue of ghetto-specific versus group-specific differences when we evaluate reported preferences for welfare.

Risks of recent welfare use also differ by family structure, as shown in figure 5.2, but mainly for women. Women reared in single-parent families have significantly higher risks of welfare receipts as adults. This generalization obtains across the three samples, with the largest differences among urban poor and Chicago mothers. The highest probability of welfare use corresponds to poor urban women reared by a single parent, among whom one in three had received welfare during the previous five years compared to approximately one in four equally poor urban women who were reared by two parents. The comparable figures for Chicago women are 28 and 23 percent, respectively. Less than 10 percent of all urban women were recent welfare recipients, yet the differences in the probability of receipt according to parental status is statistically significant. Being reared by a single parent does not affect men's propensity to use welfare, because this effect apparently is transmitted through the eligibility rules requiring custody of children and limited earnings prospects.

Exposure to welfare during childhood appears to have prolonged effects on economic well-being during adulthood. Both in Chicago and central cities across the United States, men and women reared in welfare-reliant families are more likely to receive welfare later in life, as figure 5.3 shows. The probability of adult welfare use is especially high for those reared on public assistance for most of their childhood. For example, Chicago women who were exposed to welfare during most of their childhood experienced a 31 percent probability of receiving public assistance as adults, compared to a 21–25 percent probability for those whose parents received public benefits for shorter periods or not at all. A very similar pattern obtains for all urban poor women, indicating that early experiences with poverty, not ghetto-specific circumstances, are largely responsible for the intergenerational continuity of welfare utilization. This interpretation is

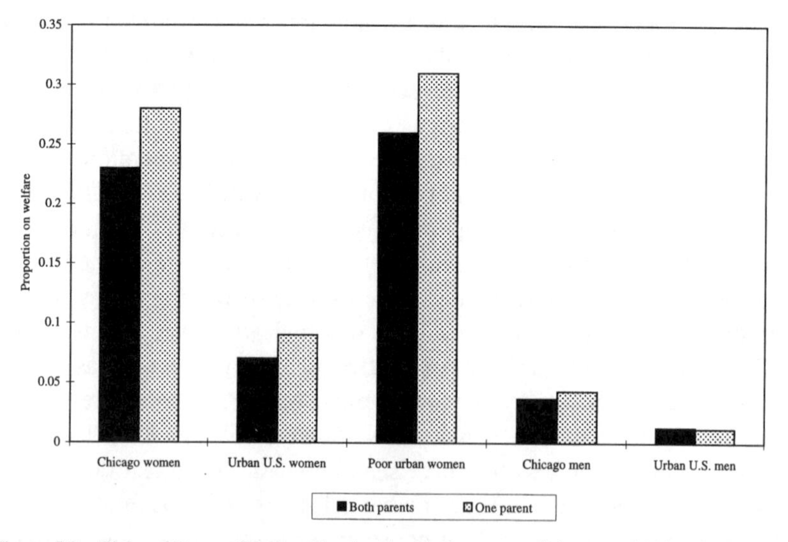

Figure 5.2 Risks of Recent Welfare Use by Family Structure: Chicago and Urban U.S. Men and Women

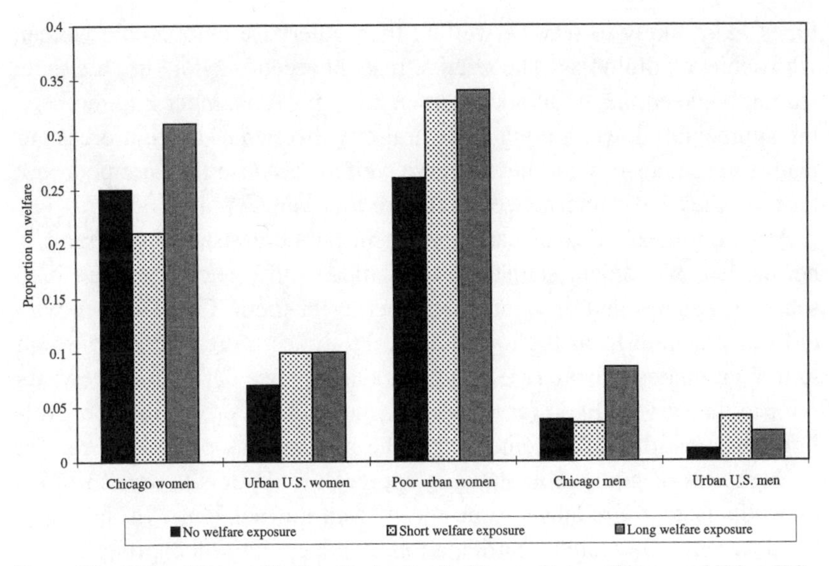

Figure 5.3 Risks of Recent Welfare Use by Family Poverty Status: Chicago and Urban U.S. Men and Women

further buttressed by evidence showing that among all urban women, welfare utilization is much lower than among the poor urban substratum or Chicago inner-city women. Yet even for these women there is evidence that early exposure to welfare is associated with higher risks of subsequent use. Also, Chicago men reared in homes where welfare use was chronic experienced an elevated probability of receiving general assistance, but for all urban men, there were no statistically significant differences in welfare use according to early exposure to public assistance. On balance, the sex differences in welfare use levels reflect the fact that eligibility for AFDC is tied to the presence of dependent children, which are typically women's responsibility.

Figures 5.4 and 5.5 show the effects on welfare participation of two major life events that the previous chapter showed to have lasting consequences on well-being in adulthood, namely, high school graduation and birth versus marriage pathway into family life. Figure 5.4, which presents the probabilities of recent welfare participation for men and women according to high school graduation status, shows that failure to complete high school is associated with increased risks of participation in public assistance programs. This relationship is consistent by sex and location, although the magnitude of the differentials varies. For example, Chicago and urban U.S. women who did not graduate from high school are 1.7

times more likely to receive welfare than otherwise comparable women who achieved diplomas. The relative risks of recent welfare use are quite similar between the urban poor women and Chicago's inner-city mothers, but appreciably lower among all central-city dwellers. This reinforces our claims that inner-city mothers receive welfare because they are poor, not because they have internalized a culture that values public doles.

Although men's risk of participating in public assistance programs is below that of women, statistically significant differences between high school graduates and dropouts also emerge for them. Chicago men who did not graduate from high school are 1.7 times more likely than high school graduates to have received general assistance during the five years prior to the survey. Even for urban U.S. men, whose probabilities of welfare participation are extremely low, the relative risk of welfare receipt by high school graduation status is appreciable. Dropouts are 3.3 times more likely than graduates to be recent welfare recipients. In the main, the differences in welfare participation by education reflect group differences in employability. That is, men and women lacking high school degrees have a more difficult time securing and maintaining jobs that pay family wages, as we document in chapter 6, and this directly affects their eligibility for and likelihood of receiving means-tested income transfers.

In the previous chapter we argued that pathways to family life—whether through marriage or birth—have long-lasting consequences for

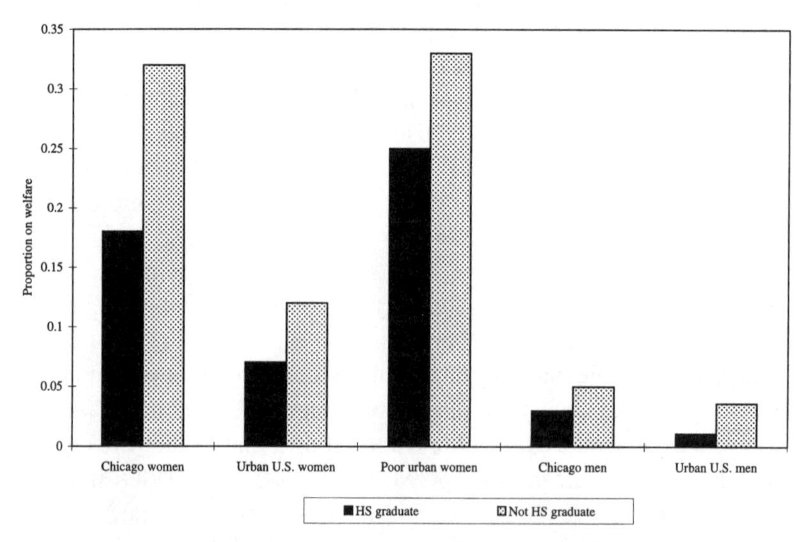

Figure 5.4 Risks of Recent Welfare Use by High School Graduation Status: Chicago and Urban U.S. Men and Women

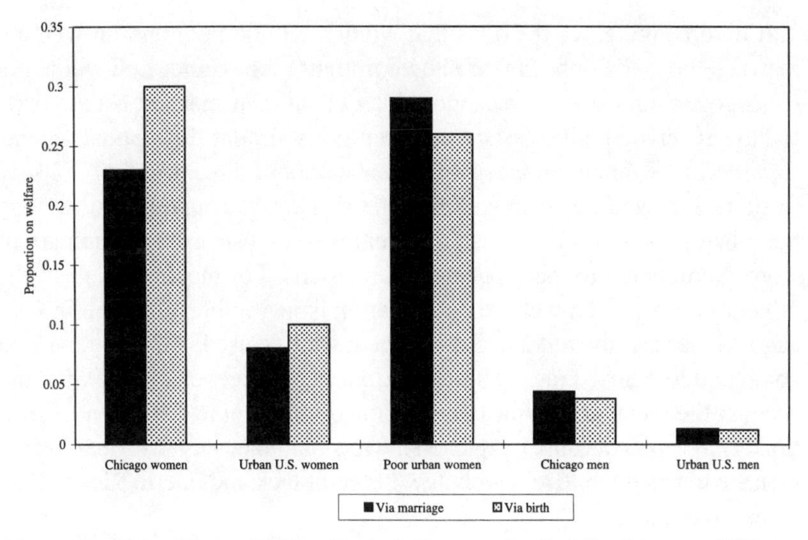

Figure 5.5 Risks of Recent Welfare Use by Pathways to Family Life: Chicago and Urban U.S. Men and Women

adult well-being. Figure 5.5 reaffirms this conclusion by showing significant differences in the risk of recent welfare use according to whether respondents began their family of procreation by having a birth or by marrying. It is noteworthy that the direct effect of pathways to family life persists even after taking into account subsequent events, which may themselves be influenced by family formation patterns. These include completing high school, getting married later in life, and labor force activity.

Among urban U.S. and Chicago women, the relative risk of recent welfare use was approximately 25–30 percent higher (or 1.25–1.3 times as likely) for those who entered family life via a birth compared to those who married before bearing a child. However, among poor women the probability of welfare use is slightly higher for those who entered family life via marriage rather than birth. Most likely this reflects the different mechanisms that draw these women into poverty following marriage rather than premarital births. That marriage is the more common pathway to family life for most poor urban women suggests that divorce, which is associated with significant income shortfalls for most women (chapter 4), is responsible for their elevated welfare participation rates. Because men are seldom the primary caretakers of their children, their pathway to family life—whether marriage or childbearing—has little bearing on their adult risk of welfare use.

If divorce increases the odds that women will be poor and on welfare, marriage does the opposite, as shown in figure 5.6. Unmarried women in Chicago are almost five times more likely than their married counterparts to have received public assistance recently. A similar differential obtains for all urban women: unmarried women are about three times more likely to be recent welfare recipients than their married counterparts. Among the urban poor, the relative risk of recent welfare use is 1.8 for unmarried women compared to their married counterparts. For most urban men, the effect of marriage on welfare participation is negligible, although in Chicago's inner city the relative risk of recent welfare use is 1.7 for unmarried compared to married men. These differences persist even after taking into account the racial and ethnic composition of the sample. This standardization is important because chapter 3 showed that inner-city, Mexican-origin men are married while relatively few of their black and Puerto Rican urban counterparts are.

Marriage and employment are the two strategies for women to exit the welfare rolls and become economically self-sufficient, and inner-city women are no different from all poor women in this regard. An analysis of AFDC welfare-leaving patterns of inner-city mothers revealed higher exit probabilities through work compared to marriage for both Puerto Rican and black mothers—the two groups with highest welfare participation rates (Tienda 1990). Because public policy has a limited role in deciding

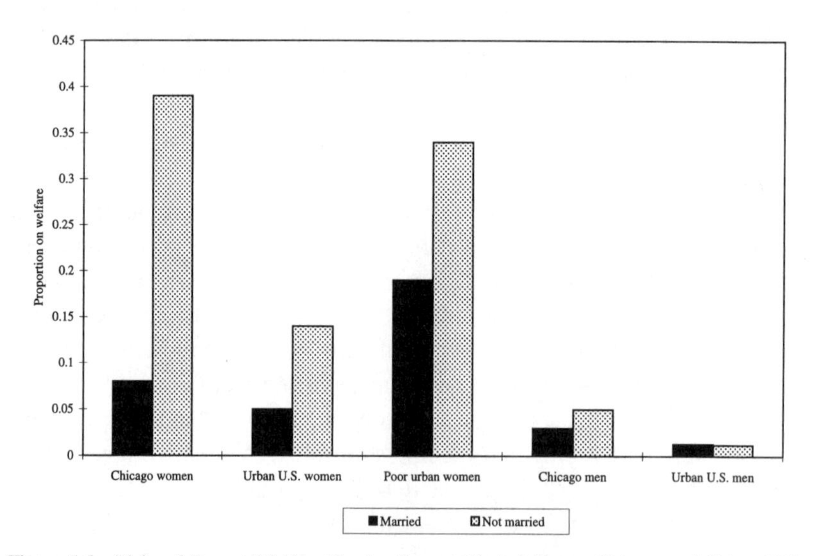

Figure 5.6 Risks of Recent Welfare Use by Current Marital Status: Chicago and Urban U.S. Men and Women

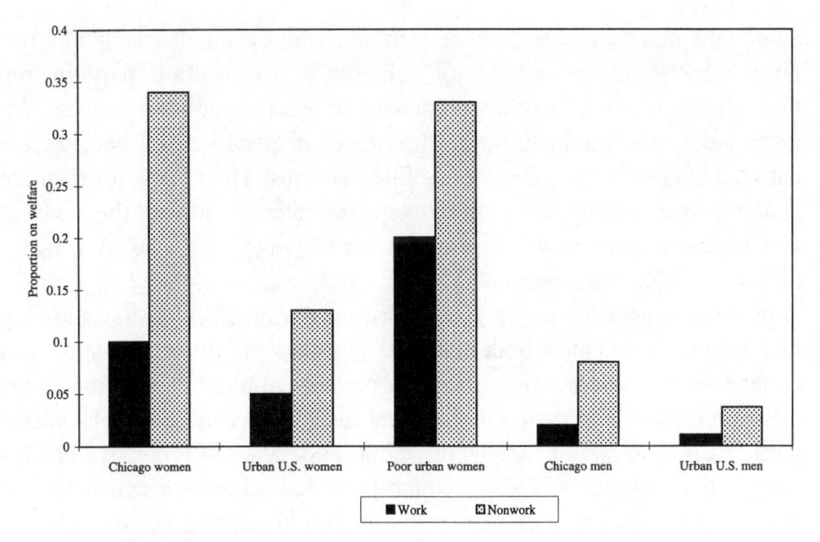

Figure 5.7 Risks of Recent Welfare Use by Employment Status: Chicago and Urban U.S. Men and Women

who will marry or divorce, attempts to influence welfare participation have focused almost exclusively on increasing employability of single parents.[2] However, the success of welfare-to-work programs is mixed, particularly among clients who require extensive training and when unskilled jobs are in limited supply.

As figure 5.7 suggests, working parents—both men and women—are less likely than jobless parents to be recent welfare recipients. The effect of work on welfare participation is especially pronounced in Chicago, where job scarcity was responsible for rising poverty during the late 1970s and throughout the 1980s. Specifically, the risk of recent welfare use is 3.5 times higher for jobless women compared to their employed counterparts. Urban U.S. women who do not work are 2.6 times more likely than those who work to be recent welfare beneficiaries, but the effect of employment status on welfare participation is somewhat smaller among urban poor women (a difference of 1.6 times). Although reasons for the smaller effect of employment on welfare participation among urban poor women are not apparent, the pattern of differentials is similar. For these women, it is conceivable that welfare participation may be more transitory than

2. Nowhere is this more evident than in the provisions of the Personal Responsibility and Work Opportunity Reconciliation Act of 1996 (PRWORA), which oblige states to move recipients into work activities while also imposing stringent time limits—both current and lifetime limits—on welfare receipt.

among the inner-city poor, whose extremely limited employment opportunities accentuate the effects of joblessness on welfare participation through chronicity rather than propensity to receive public assistance. Unfortunately, we cannot examine this issue in greater detail because the national survey lacks complete welfare histories. The results for men are generally similar, in that employment dramatically lowers the odds of welfare participation by a factor of 4 for Chicago men, and by a factor of 3.4 for urban men nationally.

In sum, figures 5.6 and 5.7 reinforce results of other studies showing that, at least for women, both marriage and work are two viable strategies either to exit welfare, conditional on participation, or to remain off the rolls altogether. For men, employment also lowers the odds of welfare participation, although receipt of public assistance is typically rare for them. This includes AFDC-U, which provided temporary assistance for married, unemployed fathers. Overall, the multivariate analysis of recent welfare participation shows that dependence on means-tested income transfers is a product of various circumstances and events that operate throughout the life course. Early disadvantages experienced during childhood influence adult lives both directly and indirectly. On the one hand, experiencing social and economic disadvantages during childhood and adolescence sets the context for behaviors that have long-term consequences for adult well-being. Poverty and scarcity of social and economic resources lead to lower levels of education and premature entry into family life. Our analyses of recent welfare participation indicate that these two turning-points in the life course have prolonged consequences for poverty and economic well-being. But poverty and early disadvantage perpetuate the experience of welfare dependence in other ways as well, either by creating additional barriers to overcoming poverty, such as confining poor families to resource-poor environments (e.g., housing projects), or by influencing attitudes toward welfare as a source of economic insurance.

An important discovery warrants emphasis. Comparisons between inner-city poor and the urban poor in general reveal more similarities than differences in the determinants of welfare participation and underscore the hardships faced by inner-city women. That the welfare behavior of urban poor women is virtually identical to that of Chicago's inner-city women suggests that poverty, rather than ghetto-specific behavior, undergirds the high welfare participation rates in concentrated poverty neighborhoods. Chicago mothers not only are worse off than urban mothers in general, on the basis of their family background and early experiences with material deprivation, but they also reside in isolated and resource-poor neighborhoods, which accentuate the disadvantages of childhood material deprivation. Again, it would appear that absence of jobs in the huge Chicago ghetto is a large part of the explanation for the high welfare participation among inner-city residents. W. J. Wilson (1987) referred to these accentuated manifestations of material deprivation as "concentration effects," which generally reflect the outcome of sorting processes that relegate minorities (especially blacks) to the most impoverished neighborhoods.

Sex differences in welfare participation are far more striking than differences by place. Women are substantially more likely to be welfare-reliant than men, largely because eligibility for AFDC depends on the presence of dependent children (under age eighteen) and because women are the primary childcare providers. For this reason, women's early family experiences influence their risk of welfare participation well into adulthood far more than is true for men. In particular, the occurrence of a premarital childbirth affects the likelihood of welfare utilization directly, by making the mothers eligible for public aid, and indirectly, by limiting their opportunities to acquire further training and full-time employment.

Thus far we have emphasized the similarities between the inner-city poor and the urban poor in general based on roughly comparable influences of family structure, educational attainment, pathways to family life, and work status. However, one noteworthy difference in the correlates of recent welfare use distinguishes the inner-city poor from the poor in general, namely, racial and ethnic differences. Whereas racial and ethnic differences in recent welfare use among all urban parents, including poor urban parents, can be attributed to group differences in characteristics that render them eligible for and predispose them to use public assistance, racial and ethnic differences persist among Chicago inner-city residents. That is, Chicago black and Puerto Rican parents who live in poor neighborhoods *are* more likely to use welfare than white men and women with similar characteristics who live in poor neighborhoods. And Mexican inner-city parents are by far less likely than all other groups to receive welfare benefits. Thus, racial and ethnic differences in recent welfare use patterns in Chicago stand in sharp contrast to all those of central-city residents and also of the urban poor.

Such results raise the possibility that inner-city minorities have developed a "taste" for welfare, possibly because of their prolonged exposure to pervasive and concentrated welfare utilization. In other words, it is conceivable that the persisting racial and ethnic differences in welfare participation among inner-city parents reflect group differences in *preferences* for welfare over work. An alternative hypothesis is that the observed group differences in welfare participation may also reflect unequal access to spatially constrained job opportunities. Put differently, the persisting racial effect on recent welfare participation observed in Chicago's inner city may reflect the discriminatory processes that produce residential segregation by race rather than race-specific welfare behavior. Specifically, inner-city black and Puerto Rican men and women may face more formidable labor market barriers than Mexican immigrants, who allegedly have been crowding out native minority workers from the low-wage job market—particularly in the context of declining employment opportunities for unskilled workers in the inner city. In addition, Mexicans—because of their immigrant composition and the likely possibility that many are undocumented—may be less likely to qualify for aid.[3] These circumstances, rather than preferences for welfare over work, may explain the persisting racial and ethnic differences in welfare participation in the inner city. If so, then group differences in participation rates should also be manifested as variation in the duration of welfare spells, that is, the chronicity of dependence.

Therefore, in the next section we address the issue of chronicity by considering whether, in fact, racial and ethnic differences in recent welfare use among Chicago parents reflect group differences in the duration of spells. Subsequently we consider direct evidence about preferences for welfare relative to work. Because information about the duration of welfare spells and preferences for welfare was not available in the National Survey of Families and Households, we restrict analyses of duration and preferences for welfare to the Chicago sample. This is not a serious limitation, however, because group differences in recent welfare use did not persist in the national samples once relevant background characteristics and early experiences were taken into account.

How Chronic Is Welfare Use?

> *Those who are stuck in a family welfare situation. Whose mothers were on welfare, and the daughter is on welfare, doesn't even think of getting a job. It's a welfare life. I not saying that 'cause I heard it on TV or anything like that. I know a few families, more than a few families, around*

3. Our ethnographic research in the survey neighborhoods revealed that a large share of recent immigrants are undocumented.

here that are like that, and they're content to live that way, to live off welfare.

Jane, thirty-three, is a white mother of one who is legally separated and appreciates what a welfare life is all about, as she indicated when queried about how family life had changed in Chicago over the past twenty years. Jane understands that it is difficult to break the vicious circle of welfare dependence. Two years of college entitles her to a teaching job at a day care center, yet an episode of general assistance gave her firsthand knowledge of life on welfare. But episodes of general assistance are mostly very short—usually providing transitional income support between jobs or following a change in marital status. Thus, her experience on "welfare" was qualitatively different from that of many inner-city parents who rely on public aid as their main source of income maintenance for protracted periods of time.

In fact, the issue of chronic dependence was at the heart of the recent debates that eventuated in the passage of the Personal Responsibility and Work Opportunities Reconciliation Act of 1996, which placed stringent term limits on means-tested income transfers—merely two years for most recipients. In some ways, the two-year time limit seems reasonable, because several studies showed that most welfare episodes are of short duration, with half of all spells lasting two years or less. However, individuals who enter a third year on public assistance experience much longer durations (O'Neill, Bassi, and Wolf 1987; Bane and Ellwood 1983; Tienda 1990; Moffitt 1992). Even though the median length of welfare receipt is less than four years (Duncan, Hill, and Hoffman 1988), and even though only a minority of all welfare recipients are long-term users, evidence of chronic dependence within or between generations has received the most attention from policy makers and the media. Although public support for income transfers as a *temporary* insurance against poverty has not totally disappeared, *prolonged* dependence has been the source of public outrage.

TANF, the public assistance program that replaced AFDC, has been dropping welfare mothers from the caseloads after two years whether or not they have alternative means of support. This feature of the program is predicated on the assumption that most welfare use is chronic, and that strict time limits were needed to combat chronic and intergenerational welfare participation. But, how long were welfare spells among inner-city parents, where economic dependence was presumably the norm rather than the exception? And how prevalent was intergenerational continuity of welfare participation? To address whether racial and ethnic differences in recent welfare participation reflect group differences in chronic

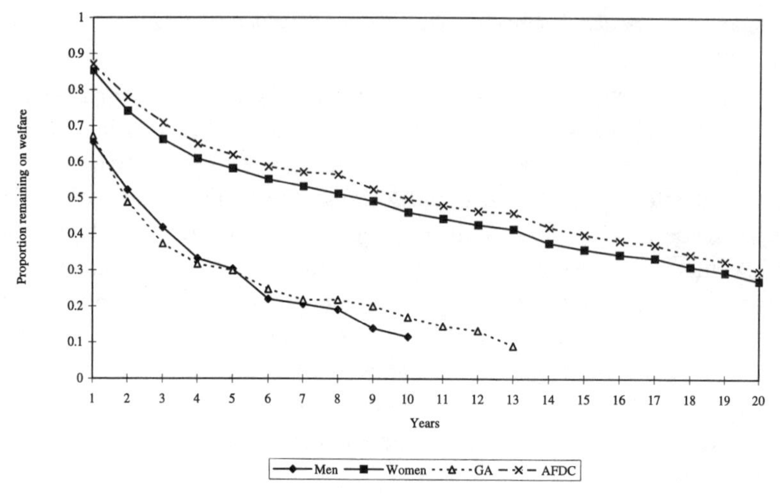

Figure 5.8 Duration of Welfare Spells by Type of Aid and Sex

dependence, we used life table methods to characterize the length of welfare spells. Appendix B reports the methodological details of the statistical procedures used to derive the welfare duration curves on which our discussion is based.

Figure 5.8 presents survivor functions, namely, the proportions remaining on a welfare grant at successive durations conditional on having been on a grant in the previous year. The survivor curves are based on respondents who ever used welfare and portray the rate of exiting the public assistance program conditional on receipt. Survivor functions depicted in figure 5.8, which compare men with women and also AFDC with GA recipients, warrant two major conclusions. First, sex differences in welfare participation correspond to program type. The AFDC survivor curve is virtually identical to the women's survivor curve; likewise, the GA survivor curve is virtually identical to men's welfare survivor curve. Second, men and women have very different welfare use experience because GA is a temporary, state-administered program, whereas AFDC eligibility and receipt required the presence of children under eighteen.

Theoretically, a single woman with a dependent child could remain on public assistance for eighteen continuous years, and even longer if additional children were born while she was on the initial grant. As primary childcare providers, women remain on welfare for longer periods than men, and they exit at a slower pace, as evidenced by the steeper slope of the male survivor functions. Obviously, chronicity is a greater problem for women than for men. Yet, our analyses of recent welfare use revealed racial effects for *both* men and women.

Less than half (47 percent) of men on AFDC-U (AFDC for unemployed fathers) remained dependent on public assistance after two years compared with 72 percent of women on AFDC.[4] Only 40 percent of men who received general assistance were still recipients after two years compared to 34 percent of women. Thus, among welfare recipients, women exit slightly faster from GA than do men, but men exit 1.5 times faster from AFDC than do women during the first two years of dependence. Probably this is due to the transitory nature of AFDC-U, which presupposes that unemployed fathers are employable and will either secure another job or return to work within a few months. Although exits from public assistance for both men and women are relatively swift during the first two years of program participation, for recipients who remain on aid for five consecutive years, the probability of exiting diminishes appreciably, especially among AFDC users. Among female AFDC users, nearly half of recipients remained on aid after nine years of consecutive use, and one-third remained dependent after ten years of welfare participation.

As expected, the chronicity of welfare use also differs by race and ethnicity. We limited the analysis of group comparison to women because of the low number of men who receive welfare benefits for an extended period of time, which precludes disaggregation by race and ethnicity. Figure 5.9 reveals similarities in the chronicity of welfare use between black and Puerto Rican women, the two groups with the longest spells and the slowest pace of exiting welfare. Within the first two years of a welfare episode, one-quarter of black and Puerto Rican women exited the rolls, compared to 30 percent of whites and 55 percent of Mexican-origin women. Half of white women who ever received public assistance exited welfare within the first five years of entering a spell. Exit rates for Mexican mothers were quite similar to those generated from national samples, in that half of Mexican mothers exited AFDC by the third year of a spell. However, for those remaining dependent more than three years, dependence is very difficult to break. Findings for Mexicans must be interpreted with some caution because of the high immigrant composition of Chicago mothers and because of the small number of spells used in estimation.

4. Although we report only survival functions for men and women and for GA and AFDC users in figure 5.8, we computed sex-specific survival curves of AFDC and GA use, on which these estimates are based.

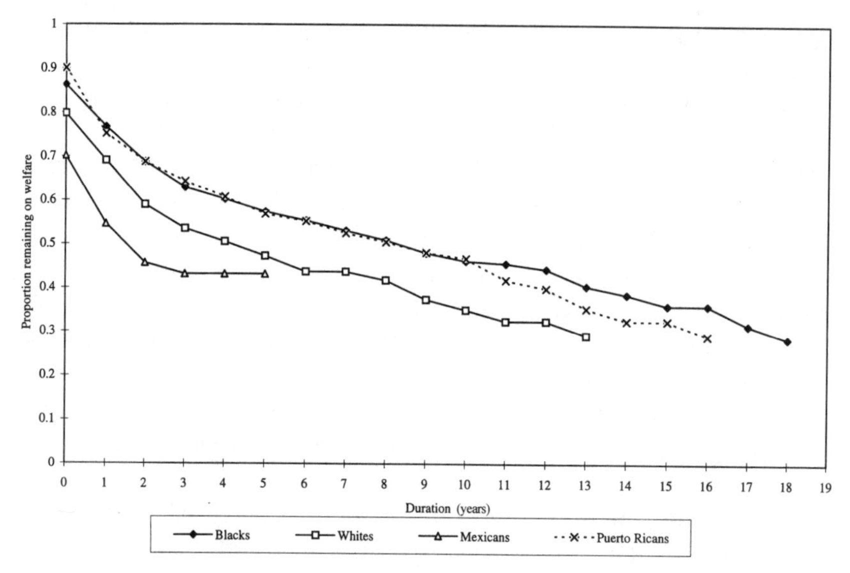

Figure 5.9 Duration of AFDC Welfare Spells: Chicago Women by Race and Ethnicity

However, these results are for women who received a grant in the first place; hence, eligibility issues are not germane. Small numbers of welfare recipients do warrant caution. After ten years of continuous welfare receipt, nearly half of black and Puerto Rican women remained on welfare compared to one-third of all white women and none of the Mexican women. On balance, the survivor functions suggest that chronic welfare use is disproportionately a minority phenomenon, but especially a black and Puerto Rican experience.

The view of prolonged dependence in Chicago's inner city, particularly among black and Puerto Rican mothers, finds additional support from summary statistics. For the pooled inner-city sample, completed AFDC spells ranged from a few months to 23 years, and incomplete spells lasted up to 26 years! The mean duration for all spells and completed spells, respectively, was 5 and 3 years.[5] A disaggregation of spells according to race and ethnicity showed the average duration of completed spells for Mexican mothers to be less than 1 year for completed spells versus 2.2 years for all spells. For them, only 14 percent of all spells in progress exceeded 4 years, and only 4 percent of completed spells lasted more than 2 years. A very different profile obtains for black and Puerto

5. Average duration for all spells is longer than for completed spells because the former includes those in progress at the time of the survey.

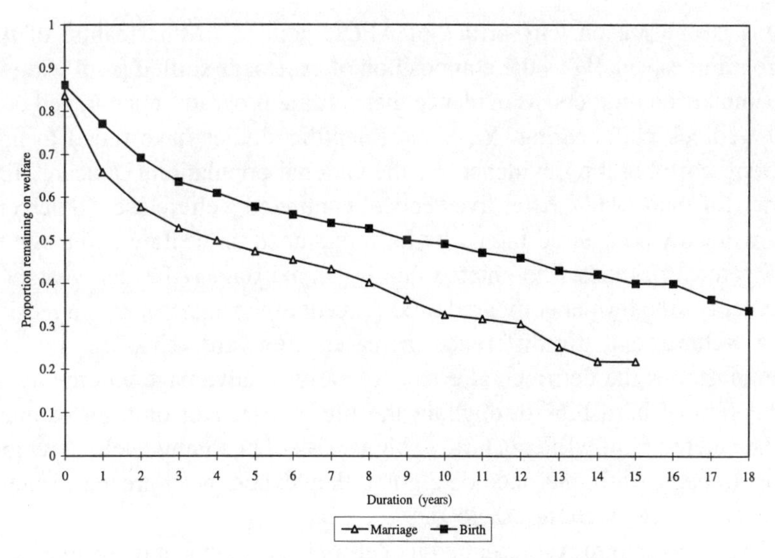

Figure 5.10 Duration of AFDC Welfare Spells: Chicago Women by Pathways to Family Life

Rican mothers. Their completed spells averaged 3 and 3.4 years, respectively, while those in progress averaged, respectively, 5.2 and 5.4 years. Moreover, the distribution of completed spells extended to 16 years for Puerto Ricans and 23 years for blacks. Of spells in progress, the highest duration (26 years) corresponded to Puerto Ricans, not blacks.

These data provide evidence of chronic dependence in Chicago's inner city, and they lend support to our interpretation that the racial and ethnic differences in recent welfare use reported in table 5.1 reflect group differences in the length of spells. Although 40 percent of all welfare episodes (including those in progress) were two years or less, roughly 18 percent of all welfare spells lasted ten or more years. Long durations were particularly common among black and Puerto Rican mothers, but not Mexican and white mothers. This information, while helpful for identifying groups at risk of chronic dependence, does not address the circumstances that determine the duration of spells.

Because eligibility for AFDC is governed by the presence of dependent children, we examined how pathways to family life affect the chronicity of welfare use. To do so, we plotted welfare duration by whether women entered family life via marriage or via birth. As figure 5.10 shows, those who entered family by birth (rather than marriage) exit welfare grants at a slower rate compared to those who entered family life via marriage.

This is one reason why critics of AFDC pointed to this feature of the program, especially as the composition of caseloads shifted from widows to unmarried mothers, as evidence that welfare programs encouraged out-of-wedlock childbearing. Yet most scientific studies have failed to find strong corroborating evidence for the national population (Duncan, Hill, and Hoffman 1988). After five years of continued welfare use, 58 percent of women who entered family by birth remained on welfare compared to 47 percent of those who entered family via marriage. After ten years, 49 percent of the former compared to 33 percent of the latter were still receiving welfare, and the difference increased afterward. Thus, figure 5.10 demonstrates the detrimental effects of early disadvantage and the accumulation of hardships throughout the life course. Not only are women who entered family life via birth at higher risk of becoming welfare recipients to begin with, but once on a grant, they experienced greater difficulties exiting the welfare caseload.

The most extreme representation of chronicity is evident in the intergenerational continuation of welfare dependence. As a special case of long durations, welfare continuity across generations was especially worrisome in the policy arena because it aroused fears that economic dependence would propagate indefinitely owing to acquired "tastes" for welfare and the development of a "welfare culture." Statistically speaking, the concern with intergenerational welfare continuity is with the strength of the correlation between mothers' and daughters' use of welfare, or more generally, whether prolonged exposure to public assistance during childhood and adolescence increases the propensity of daughters to become economically dependent on transfer payments during their reproductive years. Of course, the most dramatic cases of intergenerational welfare continuity involve daughters who receive their own welfare grants because they bear children as teenagers while they are on their mothers' grants.

The theoretical underpinnings of the statistical correlation between mothers' and daughters' welfare participation are grounded in socialization theory. Essentially this perspective implies that welfare receipt becomes statistically normative behavior with sustained exposure, even among chronic users who share mainstream values (Hannerz 1969).[6] An alternative view of welfare participation builds on a sociological tradition that documents the strong influence of family socioeconomic background on the life chances of offspring. That is, because deprived material

6. In making this statement Hannerz emphasizes that diverse behavioral patterns are situational, and that statistically normative patterns do not necessarily imply social and cultural norms.

environments attenuate the life chances of offspring, welfare mothers are more likely to produce welfare daughters than are nonwelfare mothers.

The UPFLS data provide information on whether respondents' first welfare episode occurred while they were still on their mothers' grants. Of all respondents who ever received public assistance, about 9 percent went directly from their mother's to their own grant. Racial and ethnic differences were in line with the differences in duration: more than 11 percent of black women, 9 percent of Puerto Ricans, and only 5.6 percent of white women experienced intergenerational continuity of welfare dependence because they were teenagers on their mothers' grants when they bore their first children. Based on these results, one might surmise that a welfare culture has developed in underclass neighborhoods and that inner-city minorities actually prefer welfare to work. But Hannerz (1969) would argue that chronic welfare participation is compatible with mainstream values and that the life circumstances and economic opportunities of the poor subject them to welfare dependence. A related interpretation, associated with the work of W. J. Wilson (1996, 1987) is that the observed differences may also reflect group differences in access to spatially distributed economic opportunities. To investigate whether tastes for welfare undergird racial and ethnic differences in welfare participation of inner-city parents, we examined reported preferences for work relative to public assistance.

Attitudes toward Welfare: Irrational Preferences or Limited Opportunities?

Chicago respondents were asked whether they preferred welfare to work, and we probed the reasons for their responses. In table 5.3 we present the preferences separately for the four racial and ethnic groups according to their welfare experience, that is, whether they are currently on welfare, whether they used welfare in the past, or whether they were never on welfare. Because attitudes and behaviors are intercorrelated, we expect to find higher tolerance for welfare use among those who have current or even past experience with welfare benefits.[7] Respondents could choose among three alternative answers: a job is preferred over welfare, unconditionally; a job is preferred only if medical benefits are provided as well; or welfare is preferred to work, with or without benefits.

7. Unfortunately, no comparable data is available for the NSFH. Also, we could not separate the sample by sex because of the low number of men who used welfare, either at the survey date or in the past.

Table 5.3 Preferences for Welfare versus Work by Use Status: Inner-City Parents by Race and Ethnicity (Percent)

	Blacks	Whites	Mexicans	Puerto Ricans
Current Recipient				
Job, even without med. benefits	49.7	44.6	52.3	39.3
Job, only with med. benefits	17.8	18.2	11.2	22.2
Public aid	32.5	37.2	36.4	38.5
N	438	79	41	178
Past Recipient				
Job, even without med. benefits	71.8	76.5	90.7	65.3
Job, only with med. benefits	7.8	4.1	1.9	12.7
Public aid	20.4	19.4	7.4	21.9
N	280	44	55	108
Never Recipient				
Job, even without med. benefits	78.1	80.3	78.9	74.4
Job, only with med. benefits	4.5	4.9	2.4	7.6
Public aid	17.4	14.8	18.8	18.1
N	304	238	392	168

Source: UPFLS

As reported in table 5.3, the majority of our respondents preferred work to welfare, but this expressed preference was especially pronounced for past users and for those who never used welfare. There were no significant group differences in attitudes except that Mexicans were more likely than others to prefer a job even if no medical benefits were provided. Again, this may reflect the fact that foreign-born Mexicans, who are largely unskilled and access mainly low-wage jobs in the secondary labor market, have become accustomed to jobs that offer no benefits. That over one-third of current recipients and one-fifth of past recipients (except for Mexicans) and those who never received welfare said that they preferred welfare to work raises the possibility that a welfare culture either already exists or is emerging in poor, inner-city neighborhoods. Ilana, twenty-six, a black, never-married mother of three who was on a grant when interviewed, agrees:

> *Well, the few that I probably talk to here, to my recollection, they're lazy. They're lazy. A lot of people are lazy. They don't want to work. I want to work; I do not work but I want to work. I don't want to just be on public aid. And I've talked to a lot of them: it doesn't matter to them. They say that they're going to be on aid for the rest of their life, they're never going to have anything. Like to me, you wouldn't want to live in this place here: you would want a nice place for your children to live. But the ones that I've talked to, it's like they don't care. They just don't care.*

Yet Ilana's view is at variance with reported preferences because the differences between current welfare recipients and current nonrecipients are small and statistically insignificant. Not only do reported preferences challenge the view that welfare recipients prefer public doles to work, but they also point to the importance of an adequate benefit package for working low-income mothers.

We further explore attitudes and preferences by considering responses to more general questions about welfare policy and knowledge about welfare programs, which are reported in table 5.4. Three issues were raised. First, respondents were asked to comment on whether the government

Table 5.4 Attitudes toward and Knowledge about Welfare by Use Status: Inner-City Parents by Race and Ethnicity (Percent)

	Current Recipient	Past Recipient	Never Recipient
Blacks			
Government spending on welfare			
Too much	6.1	11.0	14.1
Too little	79.5	73.0	69.8
Welfare recipients should work to receive benefits	14.6	20.1	32.5
Lack of decent job produces welfare use (agree)	87.5	71.6	68.1
N	438	280	304
Whites			
Government spending on welfare			
Too much	13.2	37.4	52.3
Too little	71.3	52.8	30.3
Welfare recipients should work to receive benefits	25.9	48.3	67.3
Lack of decent jobs produces welfare use (agree)	81.0	37.6	33.8
N	79	44	238
Mexicans			
Government spending on welfare			
Too much	29.4	50.4	65.4
Too little	46.5	21.5	15.5
Welfare recipients should work to receive benefits (agree)	31.9	43.3	66.1
Lack of decent job produces welfare use (agree)	66.4	45.8	30.0
N	41	55	392
Puerto Ricans			
Government spending on welfare			
Too much	24.1	38.8	49.9
Too little	48.2	42.8	33.1
Welfare recipients should work to receive benefits (agree)	21.7	41.8	46.9
Lack of decent jobs produces welfare use (agree)	71.2	41.4	47.1
N	178	108	168

Source: UPFLS, 1987

spends too much or too little on welfare. Not surprisingly, current welfare users are more likely to perceive the grants as insufficient compared to past users and nonrecipients. Interesting racial and ethnic differences appear in responses to this question. Among welfare recipients at the time of the survey, the majority (over 70 percent) of black and white compared to less than 50 percent of Mexican and Puerto-Rican respondents thought that the government spent too little on welfare. One-quarter of Mexican and Puerto Rican welfare users thought that the government spent too much on welfare. In fact, irrespective of their welfare use status, Mexicans were more likely than the others to report that welfare payments were too high. This attitude is consistent with their behavior, inasmuch as they are also less likely to use welfare and, as we show in the following chapter, more likely to be employed. As noted earlier, however, inner-city Mexican residents are predominantly foreign-born; thus, public assistance is a relatively novel idea for them, because no comparable programs exist in Mexico. Hence, relative to no benefits, even the meager stipends provided in Illinois seem generous to Mexican nationals. Blacks, on the other hand, were more likely than others to claim that public transfers were insufficient. They were also more likely than others to use welfare, and were keenly aware of the hardships of making ends meet with the amount of money that welfare users actually receive.

The second question pointed to the requirement of work as a condition for welfare receipt. The same pattern of relationship to welfare status reappears: welfare recipients were less likely to accept this condition, while those who were never on welfare were more likely to agree, probably because they had never experienced—nor could they imagine—the hardships associated with public aid. Most current welfare recipients disagreed with this premise. However, consistent with the reaction to the first question, blacks were more likely to oppose (only 15 percent of current users agreed) and Mexicans were more likely to agree (32 percent) that welfare recipients should work as a condition of receiving public aid. Again, this squares with their labor force experiences, as we demonstrate in the next chapter. Among past users and those who never used welfare, blacks were also the least likely to accept a work condition, while the differences among the other groups were smaller. In fact, the majority of white and Mexican nonrecipients (more than 67 percent) agreed to the work condition (compared to 47 percent of the Puerto Ricans). Note, however, that Chicago's black, inner-city residents experience the highest levels of joblessness, (W. J. Wilson 1996, 1987), as we document in the following chapter. Hence, their higher tendency to refute the work requirement largely reflects actual experiences in trying to secure work.

Last, respondents were asked to point out whether the lack of decent jobs is responsible for high rates of welfare participation. Again, the majority of current welfare recipients (more than 80 percent of blacks and whites, 71 percent of Puerto Ricans, and 66 percent of Mexicans who were welfare users) agreed that the main reason for welfare dependence is the lack of job opportunities. Among nonrecipients, support for this claim was considerably lower (around 40 percent), probably because they encountered less difficulty in finding work. Independent of their welfare-use status, two-thirds of the blacks supported the claim that the lack of decent jobs was the main cause of poverty and welfare dependence. As Irene, a never-married black mother of three, bluntly reported during the follow-up (SOS) survey,

> *Yes. They tell you to get a job, but there ain't no jobs. They say you're a fixed income, but they fail to realize that food and clothes are a lot with kids. Even someone without kids, they give them a fixed income—rent starts at $200. They give them $158 and $70 or $80 in food stamps to last them a month. How you do that? When your rent starts at $200? That's why everybody's moving to places that pay higher. And then they tell you to get a job. Where? If you get a job then, they deduct what you get from the job. You get sick, what's going to happen, then you're right back in the same hole. . . . Governor Thompson say, get a job. Where? When they build something, they building gas stations. If you work in the suburbs, you gotta have a car, then you gotta buy gas. You spending more getting to the suburbs to work then you is getting paid, so you still ain't getting nowhere. But don't nobody see that. You can't understand that if you haven't been through no trouble.*

Irene obviously understands the difficulty of exiting welfare through work because of the costs involved in holding a job, not to mention the costs of providing childcare. With only an eleventh-grade education and three children to care for, she is hard-pressed to secure a job that pays a family wage, much less health benefits. Irene is thus forced to raise her three children on the meager $386 AFDC benefit allowed for an Illinois family of four in 1987 (U.S. House 1987, 408, table 9).[8] Yet, despite the realities of life on welfare and the difficulties of making ends meet, which have been graphically described by Edin and Lein (1997), welfare mothers

8. This amount is just $44 higher than the allowance for a family of three, which indicates the monthly stipend allowed for an additional child.

became the primary target of the recent reforms that limited their lifetime eligibility for public assistance. The presumption of the recent reforms is that welfare mothers don't want to work; that employment training will enable them to qualify for jobs; and that jobs are sufficiently plentiful for all who want to work. We examine these assumptions in the following chapter.

Summary and Conclusion

> *Most of these people, they are on aid; like I'm on aid. It seems the kids grow up, and they seem like think that's the best thing to do instead of trying to go up, they're where their parents were. See, I grew up in a better neighborhood. . . . My kids are small right now, but when they grow up, I don't think I want to stay in this neighborhood. . . . It seems like people wanted to go places, they wanted to do things. . . . The kids think the same as their parents here. The neighborhood I grew up in, seems like they went for it: doctors, lawyers. This neighborhood, the kids seem happy going day by day.*

Wendy, forty, an unemployed white mother of three with a high school diploma, laments the changes in her neighborhood, which she perceives are not conducive to breaking the cycle of poverty and welfare dependence. Although she is adept in capturing the social transformation of Chicago's inner-city neighborhoods, she does not appreciate how disadvantage accumulates over the life course and limits socioeconomic options for subsequent generations. This is most dramatically represented in intergenerational continuity of welfare participation—the circumstance where mothers and daughters both depend on public aid for subsistence. Although there are instances of intergenerational welfare utilization in Chicago's inner city, and probably all impoverished central-city neighborhoods, these cases are typically rare, even in the most deprived neighborhoods.

We have described and analyzed the participation of Chicago's inner-city parents in means-tested public assistance programs, considering differences between black, white, Mexican, and Puerto Rican men and women and drawing comparisons with all central-city residents and the urban poor. In response to questions about whether welfare participation is group-specific or ghetto-specific, we find evidence that urban parents, including the urban poor, exhibit no group-specific differences in their propensity to rely on public assistance for income maintenance. Rather, average group differences in recent welfare participation rates reflect racial and ethnic differences in characteristics and early life circumstances that make them eligible for public assistance as adults. These include limited educational achievement, bearing children out of wedlock, early and prolonged experiences with poverty, and being reared in parent-absent families. Although early exposure to welfare increases the odds of economic dependence during adulthood, this intergenerational link does not necessarily imply that inner-city neighborhoods foster a culture of welfare, where public aid is preferred to work. Rather, and contrary to Wendy's view, most aid recipients indicated a preference for work over welfare, reported that welfare payments were inadequate to make ends meet, and agreed that many available jobs were inadequate to provide a better alternative to welfare because of transportation and childcare costs.

Despite the absence of racial and ethnic differences in welfare participation among central-city residents, including the urban poor, we showed that black parents in Chicago's inner city had a higher propensity than statistically similar whites or Hispanics to rely on aid for income maintenance. Also, conditional on receiving aid, black mothers exhibited the longest welfare spells and the lowest probability of exiting at any particular duration, although Puerto Rican aid recipients also experienced chronic use. These racial differences persist even after taking account of group differences in characteristics that render families eligible for public aid. Although Wendy's statement suggests the emergence of a welfare culture, our analyses showed that the highest rates of black welfare participation occurred in ghetto-poverty neighborhoods, where public housing concentrates and traps black welfare recipients in material disadvantage. As Irene and the numerous voices recorded by Edin and Lein (1997) in *Making Ends Meet* convey, once on welfare, mothers have very limited options for exiting the rolls. Thus, we conclude that the "racial effect" on welfare participation in Chicago results from discriminatory housing policy and segregation practices that confine blacks to the poorest neighborhoods, and from persisting labor market discrimination that gives whites and Mexicans preference in accessing the dwindling number of low-skilled jobs in Chicago.

This interpretation is supported by responses to the Social Opportunity Survey, where the vast majority of respondents emphasized ever more limited opportunities for earning a living on the wages provided by unskilled jobs, and asserted that minimum-wage jobs seldom provide a better alternative to public assistance, especially if they offer no health benefits. Thus, we tentatively conclude that the apparent racial and ethnic differences in welfare participation among Chicago's inner-city residents probably result from unequal employment opportunities for black, white,

Mexican, and Puerto Rican populations. In particular, these expressed preferences may point to the greater difficulty blacks and Puerto Ricans have in finding jobs that pay a living wage. We investigate this conclusion in the following chapter, which considers group differences in labor force participation throughout the life course as well as in 1987. Chapter 6 also considers more directly the evidence about welfare recipients' willingness to work by investigating whether jobless minority workers have unrealistic wage expectations, given their education and prior work experience.

CHAPTER SIX

Makin' a Living: Employment Opportunity in the Inner City

Well, I must be stereotyping, but I think the whites have better chances. Usually they're better educated, so they have more opportunities. I mean, I know some white people are on aid, but it's not, you know, a high percentage—'cause I worked for Public Aid, and we had whites, but the majority of the people were black people. And, the next was the Hispanics, and a few whites.

Like most of our respondents, Shala, a chronically unemployed black mother with a college degree, answered questions about opportunity and who is likely to get ahead with references to race and ethnicity, but she also associated having better life chances with better education. Responding to questions about opportunity and who had access to jobs, several respondents acknowledged that sometimes slots could be reserved for minorities, but there was general consensus that the best jobs go to whites, and especially educated whites, while the worst jobs go to blacks and Latinos. Another recurring theme in response to questions about what it takes to get ahead in life was the importance of adequate labor market skills. For example, when probed as to who was least likely to get ahead, Shala replied, "Hispanics, I think. They don't give them a good chance, no opportunities. I don't know [why]. I think they have the same problem blacks did—uneducated, poorly trained, poor skills."

This view about employment opportunity in Chicago during the late 1980s resonates with the tentative conclusion reached in prior chapters, namely, that racial and ethnic differences in economic well-being most likely reflect unequal employment and educational opportunities among black, white, Mexican, and Puerto Rican men and women. The precipitous decline in the number of jobs in the inner city detailed in chapter 2 further accentuated economic inequalities among Chicago's minority populations. Gloria, a black mother of two who never married and was chronically on welfare despite completing two semesters of college, told us,

> *Well, there's not a lot of jobs here. The people that are here are trying to get jobs. There's just nothing here for them, and people are just struggling to make ends meet. Even people that have jobs are catching it too. Some of them are working, and they feel the same way. They look like they're going backwards instead of forward when they should be going ahead because they are working.*

Apparently, working is a necessary condition for getting ahead economically, although low wages often limit this possibility among those who work. But Brianna, forty-one, a legally separated, black mother of two who never completed high school and was on public aid at the time of the interview, blames the rising tide of immigrants rather than the lack of jobs for the labor market difficulties of native minorities. She believes low wages result because immigrants undercut the wages of minorities. In her words,

> *Now, they don't have the jobs that they had, even back in the sixties . . . before, I don't think it's as many people are from other countries was over here like it is now, you know. And it seems to me they're sort of moving out the Americans (laughs). I mean taking up the jobs where the peoples, Americans used to have. 'Cause I had worked for quite a few in my life, I worked in a hospital until four years ago, and they laid us off. And it was a lot of foreigners over here, you know. And when they come over here, they work for little or nothing, you know what I'm saying. . . . Americans understand that they couldn't live off with what little money they were paying.*

Taken together, these voices from Chicago's inner-city parents name discrimination, inadequate labor market skills, declining employment opportunities and increased competition from immigrants as the primary circumstances responsible for the large racial and ethnic disparities in employment. Yet, outside liberal academic circles, the presumption was that the inner-city poor did not want to work and that the growth of the urban underclass was fueled by the spread of a welfare culture. Although the previous chapter concluded that inner-city parents prefer a job to welfare, self-reported preferences require more compelling evidence about the conditions under which inner-city parents *accept* a job, and about employers' willingness to hire them.

In this chapter we address this issue by examining labor force activity among parents residing in neighborhoods differentiated *both* along racial/ethnic and economic lines. Our general aim is to identify and evaluate the circumstances that produce and sustain employment inequities among blacks, whites, Mexicans, and Puerto Ricans and to determine which, if any, are unique to inner-city residents. Accordingly, we address three general questions about how color, that is, minority-group status, is related to employment opportunity, which is defined both by job availability and employer discrimination. First, does the labor force activity of inner-city parents differ from that of urban parents in general, and if so, in what ways? This question is important because it provides a benchmark against which to assess the allegedly deviant labor force behavior of inner-city residents and, in particular, the pervasiveness of joblessness among able-bodied men and women.

We find similar racial and ethnic differences in labor force activity across places, except that the *level* of joblessness is appreciably higher in Chicago compared to urban places nationally. That is, the highest rates of market activity correspond to Mexican and white workers, and the lowest to Puerto Ricans and especially blacks. Within Chicago, joblessness is more pervasive in high-poverty neighborhoods and among women. That whites and Mexicans are more likely to work than either blacks or Puerto Ricans in similarly poor neighborhoods implicates discrimination as a factor producing and maintaining racial and ethnic labor market disparities. In Chicago the observed differentials in labor force activity indicate a moderate neighborhood effect, a strong sex effect that depends on neighborhood poverty, and an especially strong racial effect. We argue that the historical and contemporary discriminatory processes that segregate blacks into the poorest neighborhoods are pertinent for understanding why living in poor neighborhoods has less deleterious consequences for the labor market activity of white and Mexican workers compared to blacks and, to a lesser extent, Puerto Ricans.

To substantiate this argument, we consider how and why racial and ethnic differences in labor market status emerge. Addressing this question requires a temporal perspective of employment experiences and explanatory frameworks that engage both individual and structural determinants of labor market success. Chapter 4 documented how and *why* racial and ethnic differences in educational attainment arise, and in this chapter we trace its consequences by examining unstable work behavior and the probability of working in a particular year, given prior life-cycle experiences and labor market opportunities. Following the theme of accumulated disadvantages developed in chapters 4 and 5, we characterize racial and ethnic disparities in labor market standing over the life course to illustrate how early underachievement produces unstable work experiences that translate into weak labor market standing among mature adults, and we examine whether the employment returns to human capital are uniform among demographic groups. The latter topic addresses the possibility of current discrimination, whereas the former addresses the possibility of prior discrimination and how early life-course choices compromised life chances during adulthood.

We demonstrate the emergence of work-experience deficits over early labor force careers that place blacks and Puerto Ricans at a disadvantage relative to Mexican and white workers. This mechanism producing labor market inequality is more accentuated among inner-city parents, especially younger cohorts, who confronted declining labor market opportunities precisely at the time they entered the labor force. Empirical analyses reveal that racial and ethnic differences in labor force participation among Chicago inner-city residents are largely driven by group differences in educational attainment and prior work experience. Once groups are statistically equalized by education, work experience, and other characteristics that influence labor force behavior, we find that racial and ethnic differences in the probability of working are greatly diminished, especially in Chicago. However, moderate racial and ethnic differences in labor force activity persist among poor urban men and women, which signal either differences in responsiveness to available labor market opportunities or more intense discrimination, or both.

A second major finding is that employment opportunities figure prominently in determining the odds of labor force activity in any given year. This finding is substantiated by large differences in labor force participation rates between Chicago and all urban places, and evidence that urban high school dropouts nationally experienced a higher probability of labor force participation than high school graduates in Chicago. Our analyses also indicate that the employment returns to human capital (education or experience) do not differ among racial and ethnic groups. The largest differences in labor force activity correspond to place, not race. Summarily stated, differences in the probability of labor force activity between Chicago residents and all urban residents are more consistent with a story about limited opportunity than with a story about racial and ethnic differences in responsiveness to job opportunities.

Finally, we examine Brianna's claim that "immigrants work for little or nothing" to resolve the paradox of high participation rates among Mexicans, the least educated group in Chicago, relative to blacks, the minority group with the highest education. The explicit message is that native minorities, like blacks and Puerto Ricans, may have unrealistic wage expectations relative to their skills and available job opportunities. Thus, the third set of questions addressed is whether discernible racial and ethnic differences exist in wages received by comparably skilled workers, whether discrimination may be responsible for observed disparities, or whether groups differ in their willingness to work. We find no evidence that black, Mexican, or Puerto Rican workers expect higher wages than white workers as a condition of accepting a job. However, we do show that, conditional on obtaining a job, minority workers receive lower wages than their equally skilled white counterparts. We conclude that, in addition to limited job opportunities, statistical discrimination colored economic opportunities in Chicago during the 1980s.

In sum, our empirical analyses address the relative importance of color and opportunity in shaping profiles of market disadvantage among inner-city residents. In keeping with the distinction between poor people and poor places, we begin by comparing the employment status distributions of Chicago's inner-city parents with all urban parents and a subsample of poor urban parents. For the inner-city sample, we consider the relative importance of place by showing how joblessness (both unemployment and nonparticipation) varies among minority groups and neighborhoods that differ in poverty density. After describing racial and ethnic differentials in labor market status, we evaluate the relative merits of four explanations in accounting for the observed disparities: (1) group differences in human capital, including both formal schooling and acquired stocks of work experience; (2) group differences in the continuity of employment; (3) group differences in "willingness" to work based on wage expectations relative to skill levels; and (4) group differences in wage discrimination. These explanations are not intended as competing alternatives because, as we demonstrate, racial and ethnic labor market inequality results from the confluence of both structural and individual circumstances that are accumulated over the life course.

Place and Racial Variation in Labor Force Activity

Recent changes in social welfare legislation send a strong message that able-bodied adults, including single women with children, must work for a living. The presumption, of course, is that all able-bodied adults are "workforce ready," and that available jobs pay living wages. In fact, most adult men do hold steady jobs, and labor force participation rates of mothers with young children have reached an all-time high (Spain and Bianchi 1996; Bianchi 1995). Nevertheless, for the United States as a whole, rates of labor force participation differ appreciably among demographic groups and according to where people live (Farley 1995).

Table 6.1, which shows large differentials in labor force status among Chicago parents and among poor urban parents relative to all urban parents by race, Hispanic origin, and prior work experiences, warrants several generalizations. *First,* average labor force participation rates are uniformly lower among Chicago parents than all urban parents.[1] Among men, unemployment rates were higher for those in Chicago compared to the rates for all urban fathers, but for women, unemployment rates were lower for those in Chicago compared to their national urban counterparts.[2] The most likely reason is that, compared to their urban counterparts, higher shares of Chicago's jobless mothers are not actively seeking employment, which is required to be classified as unemployed. Thus, although unemployment rates for Chicago mothers are generally lower than for all urban women, joblessness is higher. Similarly, men's nonparticipation rates are higher in Chicago compared to all urban fathers, owing both to higher unemployment and nonparticipation.

Second, racial and ethnic differentials in employment and unemployment rates are roughly similar across the urban populations compared, except that joblessness is more pervasive in Chicago's inner city. Among men, blacks and Puerto Ricans are extremely disadvantaged; both have high rates of joblessness stemming both from unemployment and nonparticipation, that is, joblessness that does not involve job search. Not only were employment rates of black men consistently lower than those of white men, but unemployment rates were higher by several orders of magnitude. For example, in Chicago black men were three times as likely as whites to be unemployed. Among all central cities the racial gap in unemployment was larger still—over fivefold.

Despite their relatively low education levels, Mexican men's unemployment in Chicago was lower than that of white men, 2 and 4 percent, respectively. However, at the national level, Mexican-origin men experienced higher unemployment rates than whites. These differences arise not only from the variable labor market conditions that produce the national rates, but also from the nativity composition of the Chicago sample. That is, because Chicago's inner-city Mexican population is predominantly foreign-born, and Mexican immigrants have higher labor force participa-

1. Labor force participation rates include *both* the employed and the unemployed, that is, persons who are looking for work.

2. White women are a possible exception, with unemployment rates of 3.3 percent in Chicago compared to 2.3 percent nationally. However, this difference is not likely to be statistically reliable.

Table 6.1 Employment Status Distribution of Men and Women by Race and Ethnicity: Chicago and Urban United States (Percent)

	In Labor Force		Not in Labor Force		
	Employed	Unemployed	Ever Worked	Never Worked	*N*
Chicago					
Blacks					
Men	69.7	12.9	9.1	8.3	308
Women	44.1	6.4	30.4	19.1	719
Whites					
Men	87	4.3	8.7	—	127
Women	52.6	3.3	35.5	8.6	237
Mexicans					
Men	93.1	1.7	4.3	0.9	228
Women	50.9	1.9	30.2	17	261
Puerto Ricans					
Men	76.8	7.6	12.9	2.7	148
Women	34.1	1.3	36.4	28.2	306
U.S. Central Cities					
Blacks					
Men	82.5	11.8	5.7	—	213
Women	61.9	10.2	18.2	9.8	529
Whites					
Men	93.5	2.3	3.3	0.9	613
Women	64.6	2.3	28.3	4.7	1,095
Mexicans					
Men	91.4	5	1.7	1.9	101
Women	48.5	9.5	25.1	16.9	201
Puerto Ricans					
Men	77.1	14.5	8.5	—	20
Women	27.7	5.1	28.9	38.3	62
U.S. Urban Poor					
All Men	74.6	9.8	12.2	3.4	137
Women					
Black	43.6	19.5	23	13.8	218
Nonblack	40.4	7.2	31.7	20.7	382

Source: UPFLS and NSFH

tion rates than their native-born counterparts in most labor markets (Greenwood and Tienda 1998), the aggregate participation rate for Mexicans residing in Chicago is heavily weighted by the behavior of immigrants.

Among all urban mothers, blacks participate in the labor force at higher rates than white and Hispanic women, but racial and ethnic differences in Chicago women's labor force activity deviate from the national pattern.[3] Only 44 percent of inner-city black mothers were employed when interviewed in 1987, and an additional 6 percent were looking for work. Among all urban mothers, whites were more likely to have a job in 1987 than their black counterparts, whose unemployment rates were almost two times higher (6.4 versus 3.3 percent). Hispanic mothers exhibit the lowest participation rates among the groups compared. In 1987, Mexican-origin women residing in Chicago participated in the labor force at a rate slightly below that of their white counterparts, but among all urban mothers, the participation rate of whites exceeded that of Mexicans by almost 9 percentage points. Puerto Rican women's participation rates were extremely low both in Chicago (34 percent) and all central cities (28 percent).[4]

Third, among parents who were out of the labor force, Chicago residents were far more likely than all urban parents, and even poor urban parents, to lack any prior labor market experience. Such evidence led many analysts to conclude that the majority of inner-city residents are detached from the labor market because they are unwilling to work. Especially noteworthy in this regard are the high shares of jobless black and Puerto Rican inner-city fathers who lacked any previous market experience. Approximately half of all jobless, black, inner-city fathers who were not looking for work (fully 8 percent of black, inner-city fathers) had *never* worked in the past. This compares to only 3 percent of Puerto Rican fathers who had never worked in the past. This phenomenon appears to be unique to inner-city men, because *all* urban black fathers who were jobless in 1987 reported some prior market engagement, although between 1 and 2 percent of jobless white and Mexican urban fathers reported no prior work experience.

For women, the lack of prior market experience is less surprising, because many assume family responsibilities at very young ages, but able-bodied men are expected to work if they are not enrolled in school. Black, inner-city mothers also differ from all urban black mothers with respect to their previous labor market experience. The share of jobless women lacking any prior work experience is nearly double in the inner city

3. Labor force participation rates of Hispanics based on the national survey must be interpreted with caution, particularly for men. That reported differentials accord with independent estimates based on census and survey data for this period inspires confidence in the pattern of differentials, if not in the point estimates themselves (see Bean and Tienda 1987 for differentials based on the 1980 census).

4. Urban Puerto Rican mothers included in the national survey probably represent mothers residing in New York City's inner-city neighborhoods. If so, and if neighborhood context is associated with labor market activity, then their participation rates should be very similar.

compared to all urban women—19 versus 10 percent, respectively. Chicago's inner-city white mothers differ less from their national counterparts in this respect, because four out of five inner-city white mothers who were out of the labor force at the time of the survey had worked previously. About 17 percent of Mexican women had no prior work experience when interviewed, and this share was similar in Chicago and all U.S. central cities. However, the share of Puerto Rican women with no prior work experience was appreciably higher, ranging from 28 percent in Chicago to 38 percent nationally.[5]

From these data it is not obvious whether jobless parents lack work experience because of inadequate skills; because of greater difficulties securing jobs (perhaps stemming from discrimination); because of greater competition for fewer jobs in their neighborhoods; or because they are unwilling to accept jobs offered to them. All of these explanations were voiced by Chicago's inner-city parents as reasons for unequal opportunity and unequal labor market standing, and each merits systematic scrutiny. If lack of jobs is the main problem, it is appropriate to ask why employment rates vary so much among racial and ethnic groups within a single labor market. At the national level, racial and ethnic differences may arise because groups that are regionally concentrated confront different opportunities that are spatially distributed, but this reason cannot well explain the large inequities within Chicago. Apparently inner-city white and Mexican-origin parents are more successful in securing jobs, on average, than blacks and Puerto Ricans who reside in Chicago. This is puzzling because whites and blacks, not Mexicans and Puerto Ricans, have the highest education levels. Thus, while not irrelevant, a human capital explanation is insufficient to address the racial and ethnic differences in labor market activity. Perhaps, as W. J. Wilson claimed (1996, 1987), the spatial concentration of people and jobs is more decisive than skill in matching workers to jobs, particularly when labor markets are as slack as during the 1980s and early 1990s.

As noted in chapter 2, Chicago's Mexican and Puerto Rican populations live in different neighborhoods (Casuso and Camacho 1985), and this circumstance may account for part of the disparities in labor market standing. However, differences in employment rates between them are unlikely to result strictly from job opportunities that are unequally distributed over

5. This estimate must be interpreted with caution because of the small sample of Puerto Ricans in the national survey. Nevertheless, both the Chicago and the national estimates are consistent with census-based tabulations for that period (Tienda and Jensen 1988).

space. Such an interpretation may be tenable for the national sample because Puerto Ricans are disproportionately concentrated in the Northeast, where blue-collar employment opportunities declined steeply during the 1970s and 1980s (Tienda 1989), and for mothers, who in pre-TANF days were eligible for public assistance if they did not work. As U.S. citizens, Puerto Ricans are eligible for public aid, and mothers of young children often prefer this option to low-wage work. Because 80 percent of Chicago's inner-city Mexican residents are foreign-born, they may not qualify for public assistance. However, a welfare participation story cannot explain the large racial and ethnic differences in men's labor force status in Chicago because, as shown in chapter 5, their welfare receipt rates are typically low and of short duration.

Racial and ethnic differentials in labor force activity *within a single labor market* are crucial for understanding the emergence and perpetuation of the urban underclass because they bring into focus the relative importance of color *and* spatially distributed opportunity in producing the syndrome of persisting poverty, welfare dependence, and labor market detachment. Three interpretations are plausible: (1) living in a poor neighborhood has weaker effects on the work activity of whites than of blacks; (2) Chicago's poor white neighborhoods are nowhere as poor as its black neighborhoods; or (3), some combination of these two. Table 6.2, which reveals how joblessness is spatially distributed in Chicago, addresses these possibilities.

One clear message is that employment rates decrease and unemployment rates increase, albeit not monotonically, according to neighborhood poverty level. This is neither surprising nor unexpected in light of the relationship between poverty and employment, but the large racial and ethnic differences *within* neighborhood poverty strata are noteworthy. For example, among the low-poverty neighborhoods, employment rates ranged from 95 percent for Mexican men to 83 percent for white men. This finding suggests that minority men who live in low-poverty neighborhoods are either positively selected vis-à-vis their white counterparts who live in similarly poor neighborhoods, or that similarly situated white men actually work less than minority men. Unfortunately, our data do not allow us to investigate this question.

Further, among women residing in low-poverty neighborhoods, the highest employment rates correspond to whites and the lowest to Puerto Ricans. Black women residing in low-poverty neighborhoods were employed at a rate comparable to that of Mexican-origin women, but they have relatively high unemployment rates, especially compared to their

Table 6.2 Employment and Unemployment Distribution of Chicago's Men and Women by Race, Ethnicity, and Poverty Rate (Percent)

Neighborhood Poverty Rate	Black		White		Mexican		Puerto Rican	
	Employed	Unemployed	Employed	Unemployed	Employed	Unemployed	Employed	Unemployed
Men								
<20	90.1	9.9	83.8	6.1	95.2	2.4	89.9	0
20–29	76.2	10.1	91.8	4.7	94.7	1.5	76.6	8.4
30–39	60.7	19.7	—[a]	—[a]	86.7	2.3	70.6	12.9
40+	60.6	9.9	—[a]	—[a]	—[a]	—[a]	—[a]	—[a]
Women								
<20	56.3	8.7	61.2	3	54.4	0	31.4	0
20–29	53.3	5.9	50.3	4.7	50.3	1.8	36.4	1.9
30–39	45	5	43.2	0	52.8	4.3	37.3	0.8
40+	28.1	7.9	—[a]	—[a]	—[a]	—[a]	—[a]	—[a]

Source: UPFLS

[a] Too few cases.

Hispanic and white counterparts. These unemployment differentials suggest that black women may have more difficulty finding jobs in these areas. Because these tabulations do not consider group differences in other characteristics that influence labor market behavior, such as age, education, and prior work experience—and, for women, the presence of young children—it is difficult to interpret these differentials further without a multivariate analysis, which we present below.

Neighborhoods with poverty rates between 20 and 29 percent showed great variation in employment status by race and Hispanic origin. Between 92 and 95 percent of white and Mexican men residing in such neighborhoods were employed compared to only 76 percent of similarly situated black and Puerto Rican men. For women living in neighborhoods with 20 to 29 percent poor residents, employment rates varied less by race and Hispanic origin, as approximately half were employed at the time of the survey, but only 36 percent of Puerto Rican women had a job. Black women residing in these moderate-poverty neighborhoods experience higher joblessness than their Mexican and white counterparts who lived in similarly poor neighborhoods. In these neighborhoods, black women's unemployment is only slightly higher than that of whites, but the same is true of their employment rate. Also, black and Puerto Rican men residing in neighborhoods with poverty rates between 20 and 29 percent experienced appreciably higher unemployment than their Mexican and white counterparts.

Not only are blacks and Puerto Ricans far more likely than either whites or Mexicans to reside in the poorest neighborhoods, but within these neighborhoods they experience higher levels of joblessness. Black men's employment rates in ghetto-poverty neighborhoods hovered around 60 percent. With unemployment rates of 10 to 20 percent, this means that 40 percent of black men residing in neighborhoods where over 30 percent of residents are poor do not work. Although the number of Hispanic parents residing in ghetto-poverty neighborhoods (40 percent or higher poverty) is too small for reliable estimates of employment status, among those living in neighborhoods where 30 to 39 percent of residents are poor, 87 percent are employed, and only 2 percent are looking for work. Puerto Rican men's employment rates were lower than those of Mexicans living in high-poverty neighborhoods, yet they exceeded those of blacks by 10 percentage points (70.6 versus 60.6 percent).

Similar differentials obtain for women, but in most instances Puerto Ricans fare worse than their black counterparts. Among women, black participation rates exceed those of whites in neighborhoods where less than 40 percent of all families are poor, probably because of greater economic need and the unavailability of support from spouses, but whites have the market advantage in the poorest neighborhoods. Overall, these participation differentials by race and Hispanic origin indicate a moderate neighborhood effect, a strong racial effect, and a strong sex effect, which depends on neighborhood poverty level.[6]

Many factors conspired to concentrate poverty and joblessness, including the construction of high-rise subsidized housing, the decline of high-wage manufacturing jobs, and out-migration of more affluent whites and blacks to suburban neighborhoods. Massey and Denton (1993) have convincingly argued that discrimination played a prominent role in producing and maintaining high levels of residential segregation in Chicago, the most highly segregated of all U.S. cities, and that residential segregation is a crucial mechanism restricting spatially distributed economic opportunity for blacks. Discriminatory processes that segregate blacks into the poorest neighborhoods, both historically and in times of economic contraction, may also explain why living in a poor neighborhood has less deleterious consequences for the work activity of whites and Mexicans compared to blacks and Puerto Ricans.

As W. J. Wilson (1996, 1987) has forcefully argued, however, racial and ethnic differences in labor force activity also reflect changes in the spatial distribution of jobs between the inner city and the suburbs, and these changes effectively exclude many minorities—especially blacks—from the best employment opportunities. Changes in the spatial distribution of jobs in the Chicago metropolitan area between the city and its surrounding suburbs described in chapter 2 did not go unperceived by inner-city parents. As Laura, a white, never-married mother who holds a high school diploma, notes,

> *A lot of businesses are moving out of Chicago, it seems like. Probably a government policy matter. Like they say, "We're going to tax this, and we're going to tax that," and they're going to drive the companies away! We got to do something to keep them here. They say that "Chicago works!" Well, I don't think it's true! Chicago don't work.*

Carol, a married, white mother of two who has been steadily employed except for maternity-related absence, appreciates the disparity in job opportunities available in Chicago and its surrounding suburbs. Despite her

6. How neighborhoods came to be stratified along racial, ethnic, and income lines is beyond the scope of our analysis, but has been discussed eloquently by Massey and Denton (1993). The differentials in labor force activity are but one manifestation of these deeper causal processes.

success in keeping a low-wage job (she had no welfare history, but qualified for and received food stamps), she argues for policies to keep jobs in the city.

> *Oh . . . I think they should try to bring more businesses in, keep more businesses in the city. The suburbs are experiencing a fantastic growth. I mean, when I go out there to visit my friends, it just astounds me how much that empty land is being built on, and unfortunately, people down in the city cannot get to those jobs easily. They don't have cars . . . and, you know, transportation goes only so far, public transportation. And I know there's jobs there and out there but . . . I don't know. So probably to keep the businesses here and try to improve transportation to make it easier for people to go outside the city and work. Instead of always trying to funnel them into the city.*

This policy would be especially beneficial for blacks, whose options have been more restricted by the legacy of residential segregation (Drake and Cayton 1993; Massey and Denton 1993). Mirroring the racial and ethnic differences in labor market status described above, as well as the words of Laura and Carol, inner-city parents' perceptions of changes in the distribution of employment opportunity differed systematically along color lines, as shown in table 6.3.

When asked whether their job opportunities were better in the city or the suburbs, nearly three-fifths of black fathers and over half of black mothers reported that the best job options were in the suburbs, but these perceptions differ according to neighborhood poverty levels.[7] Black men and women saw their best job opportunities outside the city limits, particularly if they resided in a neighborhood where over 30 percent of residents were poor. In fact, only one-third of black fathers from ghetto-poverty neighborhoods indicated that the best job opportunities were in the city, compared to over 60 percent of similarly situated white and Puerto Rican fathers, and about half of Mexican fathers. Likewise, black women, but especially those residing in mixed-poverty neighborhoods, perceived more limited job options in the city than did their white, Mexican, and Puerto Rican counterparts. Among black mothers, just over two in five reported that their best job prospects were in the city, compared to nearly three-fourths of white and Puerto Rican women and almost three-fifths of Mexican-origin mothers. A similar pattern of differentials obtains for mothers residing in ghetto-poverty neighborhoods. These perceptions

7. Item 47, the last in the current employment section, was worded, "Do you think the job opportunities for you are better in the city or in the suburbs?" The fixed response choices included the following: city; suburbs; same (if volunteered); and don't know.

Table 6.3 Perceptions of Job Opportunities by Current Neighborhood Poverty: Chicago Men and Women by Race and Ethnicity (Percent)

	Neighborhood Poverty							
	<30% Poor				≥30% Poor			
Best Jobs in:	City	Suburb	Same	*N*	City	Suburb	Same	*N*
Men	50.8	44.6	4.6	532	39.6	56.3	4.1	279
Black	45	52.2	2.8	166	34.5	62.7	2.8	142
White	68.5	23.3	8.2	97	64.2	26.1	9.7	30
Mexican	53.1	41.1	5.9	176	53.6	37.5	8.9	52
Puerto Rican	62.3	29.5	8.1	93	60.9	31.4	7.7	55
Women	52.2	44.9	2.9	893	52.3	46.1	1.5	630
Black	43.4	55.6	0.9	326	49.3	49.5	1.3	393
White	75.1	18.4	60.4	179	67.6	24.7	7.7	58
Mexican	58	35.1	6.9	211	69	28.9	2	50
Puerto Rican	72.5	24.5	3	177	70.5	27.7	1.9	129

Source: UPFLS

square with higher levels of joblessness experienced by blacks. Taken at face value, these responses suggest that black parents have fewer opportunities in the city than do whites, Mexicans, and Puerto Ricans, and they parallel the racial and ethnic differences in labor force status reported in table 6.1.

In summary, the descriptive overview revealed that inner-city blacks are worse off than urban blacks nationally with respect to their labor market standing, but this is not so for whites or Mexicans. Puerto Ricans are worse off than Mexicans even when they reside in the same city, or in equally impoverished neighborhoods. Blacks and Puerto Ricans may be worse off than whites and Mexicans because they tend to live in poorer neighborhoods, but they may live in poorer neighborhoods precisely because of the more intense housing and labor market discrimination they experience.[8] How much of the poor labor market standing of blacks and Puerto Ricans can be traced to group differences in skills (human capital endowments); how much to changes in the demand for unskilled workers (opportunities); how much to values that denigrate work and render welfare an acceptable means of support (willingness to work); and how much to discrimination (color) requires more systematic scrutiny. In the following section we outline and evaluate these various explanations for racial and ethnic inequality among urban parents nationally and Chicago inner-city parents specifically.[9]

Human Capital and Labor Market Inequality

There's a lot of people out there that can't get ahead. No matter how hard they try, they can't get ahead. Because a lot of people don't have the education to go back to do the things that they wish that they could do.

Susan, a divorced, white mother of three who is thirty-four, has completed only eight years of school and is chronically welfare-reliant. She

8. Causality is difficult to establish, but the process is surely self-perpetuating, as described by Massey and Denton's account of residential segregation and the perpetuation of the urban underclass. This occurs when, for example, welfare benefits include subsidized public housing. In Chicago, the vast majority of subsidized housing consists of high-rise projects in racially segregated neighborhoods.

9. Obviously, there is no single explanation for racial and ethnic disparities in socioeconomic outcomes, but W. J. Wilson's emphasis on structural forces as the main reason for the rise of concentrated poverty during the 1980s requires additional supporting evidence. The UPFLS of Chicago is particularly well-suited to address his thesis because the social transformation of Chicago's inner-city neighborhoods was the genesis for his ideas.

provides a forceful argument for the disadvantaged labor market standing of minority workers, namely, that they lack adequate skills to compete for the available jobs. Simply stated, relative to the native white population, minorities have failed to keep pace with the rising educational requirements of a high-technology service economy. We provided some support for the skills mismatch argument in chapter 2. Work experience, another form of human capital that is accumulated over the life course, also enhances workers' attractiveness to employers, but this presupposes that workers have equal opportunity to acquire it in the first place. As we illustrate below, racial and ethnic differences in labor market standing are developed over the life course as unstable job experiences eventuate into accumulated disadvantages and provide negative signals to employers about the employability of some inner-city workers.

Several questions are relevant for our interest in racial and ethnic labor market inequality from a human capital perspective. First, how do racial and ethnic differences in human capital arise over the life course? And are racial and ethnic differences in educational attainment and experience consistent with labor market standing? That is, on average, do the most highly educated workers and those with the most labor market experience have the best labor market outcomes? Second, do racial and ethnic groups differ in their willingness to work? Third, are the employment and wage returns to education and experience uniform among demographic groups?

Undoubtedly, the decline of unskilled and semiskilled jobs in decaying urban centers exacerbated the economic plight of inner-city residents during the 1980s and early 1990s. Because blacks were disproportionately represented in such jobs, their precipitous decline partly explains why blacks seem to have fared worse in the labor market than their white counterparts. However, the industrial employment trends outlined in chapter 2 cannot easily explain racial and ethnic inequality within a single labor market. Especially difficult to reconcile from either a human capital or structural perspective is why poorly educated, recent immigrants to Chicago, who often lack fluency in English, appear to be more successful at securing jobs in Chicago than U.S.-born blacks. Conceivably discrimination operates to produce these egregious differentials, apparently punishing blacks who play by the rules and strive for high school education. However, this explanation is incomplete without considering pre-market discrimination as well as competition from immigrants who allegedly undercut wages of native workers. Also relevant for understanding racial and ethnic differences in current market standing are variable preferences for work over welfare, which themselves are the result of prior opportunities to work and earlier experiences with discrimination.

In practice, it is difficult to quantify the relative import of human capital and discrimination in producing racial and ethnic profiles of labor market inequality because current outcomes reflect the legacy of past discrimination and because groups may differ in their willingness to work, especially if they become discouraged after prolonged but fruitless job searches. Past discrimination will be reflected in different levels of accrued labor market experience among racial and ethnic groups, as well as unequal educational opportunities. Human capital (education, training, and work experience) accumulates over the life cycle, but so too do market disadvantages in the form of frequent jobless spells, the absence of training opportunities, and difficulties in securing job offers. Current discrimination produces disparate outcomes among comparably skilled and experienced groups who confront similar opportunities. That is, if black parents are less successful at finding jobs or are paid less than *comparably educated* whites or Mexicans, then discrimination can be invoked as one reason for racial and ethnic differences in labor market outcomes.[10] We belabor this point because the dominance of structural explanations for the rise of concentrated joblessness among inner-city minority populations has deflected attention from the past and continuing role of racial discrimination in generating racial and ethnic labor market inequality in both good and bad economic times.

Group differences in early work experiences may accentuate current labor market inequality for other reasons. A history of employment instability and repeated spells of joblessness could provide a negative signal to prospective employers about workers' reliability and willingness to work. From the workers' perspective, frequent spells of joblessness foster discouragement and eventual labor market withdrawal, but high turnover rates also produce unreliable income flows and perpetuate poverty. Among the employed, minorities may be more vulnerable to job loss than employed whites because they are more likely to work in smaller firms that are susceptible to closing or because white workers—even those with less seniority—often are preferred to minority workers (Hodge 1973; Tienda 1989). Therefore, we also consider whether minority parents are more vulnerable to job loss and, if so, whether this results largely from "structural factors" (e.g., plant shutdowns, shop closings, and the like),

10. Of course, residual explanations are less compelling than direct evidence, such as that derived from hiring audits, but such information is seldom available.

or whether voluntary departures and dismissals are largely responsible for racial and ethnic differences in labor market separations.

A third common explanation for the disadvantaged labor market status of inner-city minority groups focuses on alleged differences in willingness to work. Self-reported indicators of willingness to work are not entirely convincing, but empirical evidence that minority populations are willing to accept lower wages than comparably skilled whites in order to secure a job would be more persuasive. To evaluate inner-city parents' self-reported willingness to work in a more rigorous way than in chapter 5, we compare reported minimum working conditions, including wages, of comparably skilled and experienced workers. Evidence that minority parents expected higher wages than nonminority parents with similar stocks of human capital would challenge the salience of discrimination as a mechanism producing racial and ethnic labor market inequality.

In what follows we first demonstrate how group differences in early life-course experiences are accentuated through labor force careers that result in differentially lower stocks of work experience. This mechanism producing labor market inequality is not unique to inner-city parents, but it is accentuated among them because of the more limited employment opportunities confronted by younger compared to older cohorts. Subsequently we consider whether the employment returns to human capital are similar for black, white, Mexican, and Puerto Rican parents, and how minority-group status in turn influences workers' employment stability. Comparisons between Chicago and the national sample help isolate the importance of place-specific circumstances in generating profiles of disadvantage. Finally, after demonstrating that inner-city minority parents do not have unreasonable wage expectations relative to their human capital endowments, we argue that discrimination influences racial and ethnic differences in labor market inequality more than differences in labor market attachment. Our focus on how group membership circumscribes labor market opportunity *throughout the life course* explicitly acknowledges that, in addition to changes in labor demand and group differences in human capital, differential treatment of blacks, whites, Mexicans, and Puerto Ricans continues to produce and maintain racial and ethnic labor market inequality.

Accumulating Disadvantages through Human Capital

Chapter 3, which documented group differences in years of school completed, showed Mexicans to be the most educationally disadvantaged, and chapter 4 identified family structure and early experiences of poverty as two key mechanisms perpetuating educational disadvantages intergenerationally. In this section we analyze work histories of urban and inner-city parents to illustrate two mechanisms that produce racial and ethnic labor market inequality, namely, the accumulation of work experience and human capital.

Prior chapters identify several reasons to expect different age-experience profiles between Chicago parents and all urban parents. These include (1) unequal opportunities to work; (2) differential responsiveness to employment opportunities; and (3) racial and ethnic differences in normative expectations about market roles of men and women. Tables 6.4 and 6.5 illustrate how work-experience deficits accumulate over the life course of parents and their inner-city racial and ethnic counterparts. From the workforce histories of both the National Survey of Families and Households and the Urban Poverty and Family Life Survey, we computed three summary measures that depict employment instability and work experience over the life course. These include (1) the total number of years worked by five-year age groups; (2) the share of time spent at work after age eighteen; and (3) the total number of jobs held, including the current job.[11] These summary measures provide valuable information about work careers, but the age-specific summary comparisons do not reveal the actual life-course trajectory of any individual parent. Rather, they approximate the work careers of successive age cohorts of parents. It is also crucial to stress that the various age cohorts experienced quite different labor market conditions because of changes in the demand for unskilled labor after 1974 (see chapter 2). Still, comparisons based on the static cohort measures provide strong benchmarks about the nature and magnitude of life-course differentials in employment experiences.

These results warrant several generalizations about racial, ethnic, and gender differences in the work careers of urban and inner-city parents. First, relative to urban parents nationally, Chicago's inner-city mothers and fathers accumulated a larger work experience deficit over their adult life course than did all urban parents. Although women work fewer years than do men at every age group, the sex gap in work experience is smaller

11. The total number of jobs refers to the entire work history and thus includes jobs held prior to age eighteen. However, measures of the share of time at work must be based on a fixed age referent. We use age eighteen, the minimum age for eligibility in the survey and the modal age for completion of high school, to minimize potentially confounding effects of racial and ethnic differences in the timing of first labor market entry.

Table 6.4 Cumulative Proportion of Time at Work Since Age 18 and Mean Number of Jobs Held: Chicago and Urban U.S. Women by Race and Ethnicity

	Chicago					Urban United States				
Ethnicity	18–24	25–29	30–34	35–39	40+	18–24	25–29	30–34	35–39	40+
All Women [*N*]	[259]	[343]	[355]	[292]	[274]	[339]	[513]	[614]	[546]	[398]
Years worked	1.4	2.9	5.6	8.8	9.8	2.1	5.2	8.0	11.1	13.8
% Time worked since 18	26.1	28.1	38.6	44.9	39.2	42.0	51.2	55.0	57.0	56.5
# Jobs held[a]	1.0	1.4	2.1	2.5	2.5	1.1	1.5	1.7	1.9	2.1
Blacks [*N*]	[135]	[162]	[161]	[136]	[125]	[93]	[115]	[146]	[108]	[70]
Years worked	1.4	2.7	5.5	9.2	10.8	1.6	4.7	7.9	13.0	15.7
% Time worked since 18	26.3	26.6	38.1	47.2	43.3	34.0	45.5	56.3	65.6	64.9
# Jobs held[a]	1.0	1.3	2.1	2.6	2.6	0.9	1.3	1.5	1.7	1.8
Whites [*N*]	[27]	[41]	[60]	[52]	[57]	[218]	[359]	[417]	[393]	[297]
Years worked since 18	1.3	3.3	7.2	12.0	9.3	2.4	5.6	8.1	11.2	14.0
% Time worked since 18	15.2	29.2	49.2	60.3	36.8	49.5	57.0	55.3	58.4	56.8
# Jobs held[b]	1.5	2.0	2.9	3.0	2.9	1.2	1.5	1.8	2.0	2.2
Mexicans [*N*]	[50]	[59]	[63]	[51]	[38]	[20]	[36]	[40]	[34]	[24]
Years worked since 18	1.5	3.7	6.0	7.3	6.5	2.2	4.2	8.5	8.1	11.1
% Time worked since 18	26.4	35.6	40.3	34.8	25.7	35.6	35.9	56.0	40.0	44.9
# Jobs held[a]	1.1	1.6	1.7	2.1	1.5	1.1	1.6	1.3	1.3	1.6
Puerto Ricans [*N*]	[47]	[81]	[71]	[53]	[54]	—[b]	—	—	—	—
Years worked since 18	1.8	3.0	4.7	4.8	6.7					
% Time worked since 18	28.7	27.7	29.8	24.8	27.0					
# Jobs held[a]	1.1	1.4	1.7	1.8	1.9					

Source: UPFLS and NSFH

[a] Includes jobs held concurrently.

[b] Puerto Rican sample too small for reliable estimates.

Table 6.5 Cumulative Proportion of Time at Work Since Age 18 and Mean Number of Jobs Held: Chicago and Urban U.S. Men by Race and Ethnicity

	Chicago					Urban United States				
Ethnicity	18–24	25–29	30–34	35–39	40+	18–24	25–29	30–34	35–39	40+
All Men [*N*]	[84]	[171]	[194]	[188]	[174]	[92]	[209]	[326]	[317]	[269]
Years worked	1.8	5.8	9.0	12.5	17.1	4.9	8.6	11.7	16.0	21.3
% Time worked since 18	38.1	55.5	60.7	63.7	67.4	79.9	81.7	78.4	80.8	86.0
# Jobs held[a]	1.4	2.2	2.9	3.1	3.2	1.2	1.4	1.5	1.6	1.8
Blacks [*N*]	[51]	[65]	[66]	[60]	[66]	[23]	[36]	[57]	[56]	[45]
Years worked	1.5	4.7	8	11.8	17.8	4.4	7.9	11	15.5	20.4
% Time worked since 18	33.2	45.6	54.4	60.4	70.4	72.5	82.2	73.9	78.4	84
# Jobs held[a]	1.3	2.1	2.8	2.8	3.1	1	1.2	1.7	2	1.5
Whites [*N*]	—[b]	[18]	[34]	[37]	[31]	[61]	[155]	[232]	[240]	[209]
Years worked	—	7.9	10.4	14.0	18.0	5.2	8.6	11.5	16.0	21.1
% Time worked since 18	—	73.0	67.2	70.9	69.2	85.1	80.3	77.8	81.0	86.1
# Jobs held[a]	—	3.1	3.2	3.8	4.7	1.2	1.5	1.4	1.6	1.4
Mexicans [*N*]	[14]	[49]	[69]	[49]	[47]	—[b]	[16]	[34]	[13]	[14]
Years worked	4.3	8.3	10.6	13.9	14.2	—	9.6	14.2	17.9	24.2
% Time worked since 18	77.2	76.7	71.7	68.8	57.1	—	85.0	88.9	85.8	88.5
# Jobs held[a]	2	2.3	2.9	3.5	3.0	—	1.2	1.5	1.6	1.9
Puerto Ricans [*N*]	[12]	[39]	[25]	[53]	[30]	—[c]	—	—	—	—
Years worked	2.4	5.5	9.2	12.3	19.5					
% Time worked since 18	52.3	58.5	62.0	63.2	76.2					
# Jobs held[a]	2	2.5	3.1	3.0	3.4					

Source: UPFLS and NSFH

[a] Includes jobs held concurrently.

[b] Unweighted cell size less than 10.

[c] Puerto Rican sample too small for reliable estimates.

among inner-city parents than among all urban parents. For example, by age twenty-four, Chicago mothers averaged one and one-half years and fathers averaged two years of labor market experience, respectively. By comparison, all urban fathers averaged five years of work experience, and mothers slightly over two years by the same age. Urban fathers accumulated approximately sixteen years of market experience between their fortieth and forty-fourth birthdays, but the comparable experience stock averaged by Chicago's inner-city fathers was only twelve to thirteen years. This is a fairly substantial experience gap for the older cohort who was unlikely to have entered the labor market during a period of declining opportunities. By middle age, the experience gap between all urban fathers and their inner-city counterparts was about three and one-half years. Inner-city mothers acquired approximately ten years of market experience by ages forty to forty-four, compared to fourteen years for all urban mothers. Thus, by age forty-four, the average experience gap of inner-city mothers relative to all urban mothers was comparable to the respective gap for fathers—approximately four years.

A second generalization is that inner-city parents experience greater employment instability than all urban parents. This is evident in the average number of jobs held by specific ages and the share of time worked by various stages of the life course, but the differentials are more pronounced for men than for women. Although Chicago's fathers aged forty or more worked an average of seven of every ten years following their eighteenth birthday, all urban fathers worked approximately eight of every ten years. Moreover, Chicago inner-city fathers acquired their lower stocks of work experience over a greater number of jobs compared to urban fathers nationally: 3.2 versus 1.8, respectively. In addition to signaling greater employment instability among Chicago compared to all urban fathers, this means that the latter acquire more job-specific experience, which is more valuable to employers than general market experience (Becker 1964). Similar differentials in employment instability obtain between urban and inner-city mothers, but owing to the higher incidence of voluntary intermittence among women, the cumulative number of jobs held by Chicago and all urban mothers are relatively similar—2.5 versus 2.1 jobs by age forty-five.

A third message is that the gender gap in the acquisition of labor market experience begins early and grows throughout the life course. At the national level, urban fathers accrued 2.8 more years of work experience than mothers did by age twenty-four. Women's experience deficit relative to men's reached 7.5 years by ages forty to forty-four. The gender gap in labor market experience is similar between Chicago parents and all urban parents, with the noteworthy caveats that the average number of years worked by age groups is significantly lower in the inner city (see point 1 above) and that the gender gap in experience is smaller among inner-city parents compared to urban parents nationally, particularly at lower ages. This results primarily because of the low participation rates of young inner-city men rather than the high labor force activity rates of mothers.

Fourth, racial and ethnic differentials in accumulated labor force experience are highly variable by age, reflecting changes in job opportunities confronted by younger and older cohorts (see chapter 2).[12] Nevertheless, it appears that labor market hardships associated with industrial restructuring were not experienced uniformly among demographic groups, even within a single labor market. Comparisons between the national urban sample and the Chicago inner-city sample suggest unique influences of race and place in structuring labor market experiences. For example, by ages forty to forty-four, urban mothers acquire approximately five fewer years of work experience than their racial-ethnic counterparts residing in Chicago. Among inner-city mothers, the largest experience deficit at these ages vis-à-vis white women corresponds to Mexican-origin—not black—women. By contrast, among all urban mothers, the racial gap in work experience was negligible throughout the life course. Relative to their urban racial and ethnic counterparts, inner-city fathers ages forty to forty-four accumulated experience deficits ranging from 2.6 years for blacks to 10 years for Mexicans. Oddly, inner-city Mexican fathers exhibited the largest experience deficit vis-à-vis whites, but this is a statistical artifact of excluding employment episodes in Mexico from the work histories of these predominantly immigrant men.[13]

Among whites and Mexicans, the cumulative work experience measures also indicate greater instability among inner-city compared to urban residents nationally, especially among whites. White, inner-city fathers aged thirty-five and over worked approximately 70 percent of the time

12. Unfortunately, the sample size limitations of the national urban sample preclude parallel race and ethnic disaggregation because the Puerto Rican samples and the Mexican samples of young men and women are too small for reliable estimates.

13. The convergence in experience profiles of Mexican and white men must be interpreted with caution because we discovered a systematic reporting bias among Mexican men, many of whom excluded their employment history in Mexico. Because the Mexican inner-city sample is disproportionately foreign-born, the extent of underreporting is potentially quite serious. Therefore, our estimates of Chicago Mexican men's cumulative work experience are extremely conservative, particularly among older ages.

after age eighteen, but their work experience was acquired over four to five jobs. At the national level, urban white fathers acquired more labor market experience by working 83 to 86 percent of the time after age eighteen, and they distributed this work experience over fewer jobs, 1.3 to 1.5, on average. Thus, relative to their national counterparts, inner-city fathers not only experience greater employment instability and accrue fewer years of work experience, but also have lower stocks of job-specific tenure, which, other things equal, is more valuable to employers than general experience.

In sum, the tabular results depicting the diverse work histories of inner-city and urban parents shed light on the great variance in current labor force status reported in table 6.1. But descriptive tabulations showing large demographic variation in early work experiences cannot establish whether racial and ethnic differences in current employment status persist among groups with comparable stocks of human capital who reside in the same labor market. Put differently, it is conceivable that racial disparities in joblessness will be trivial among comparably skilled and experienced parents. However, it is also possible that education and work experience do not have the same exchange value for minority and nonminority parents. Therefore, we next examine whether the employment returns to education and experience differ along racial and ethnic lines. Because much of the recent literature on inner-city joblessness has emphasized the importance of structural factors in generating labor market inequality, for the Chicago sample we also consider how changing economic circumstances influence job separation.

Multivariate Analyses of Labor Force Participation

We analyzed a person-year file using logistic regression to address whether racial and ethnic differences in labor force participation are attributable to differences in human capital stocks. The person-year file considered all work episodes from age eighteen to the survey date. Correlates of labor force activity include high school graduation status and actual years of work experience, as well as selected family background characteristics (i.e., mother's education, family poverty status, and family structure) and life-course experiences (events) and circumstances known to affect work behavior (i.e., marital status, the presence of pre-school-aged children, and age). As in prior chapters, appendix B reports on the technical and model-specific details for the findings that follow. Results for the logistic regression analyses are presented in tables B6.2 (for men) and

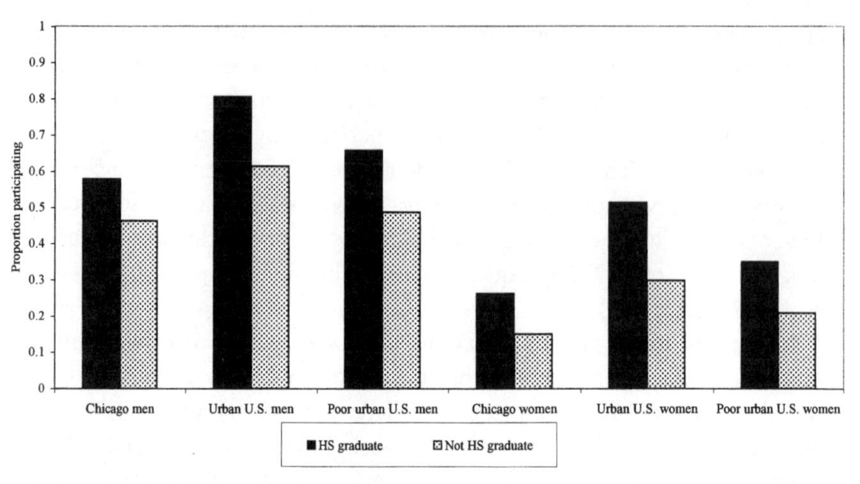

Figure 6.1 High School Completion Effects on Labor Force Participation: Chicago and Urban U.S. Men and Women

B6.3 (for women), but for ease of interpretation, coefficients were converted to probabilities and presented graphically in figures 6.1 through 6.3.[14] Thus, the bar graphs represent adjusted probabilities of labor force participation by high school graduation status, work experience, and minority-group status for men and women in each of the three samples compared. Because family background transmits its influence most directly on early investments, notably educational attainment and early work experience, the direct influence on labor force activity of being reared in a parent-absent or a poor family is generally weak. Therefore, we do not present these results in graphic form.

The most striking finding in figure 6.1 is the low level of labor market activity in Chicago compared to urban areas nationally. Acquisition of a high school degree or more significantly increases the probability of both men's and women's labor force participation, but the likelihood that people will work in any given year depends on where they live. For women, the odds of labor force participation were twice as great among all urban women compared to Chicago inner-city women, and these odds were uniform by education level. Among men, Chicago residents were only three-fourths as likely to work in any given year compared to their urban national counterparts. Chicago male high school graduates experienced a 58 percent probability of labor force participation (either employed or

14. Appendix Table B6.4 reports the covariate values used to derive these probabilities.

looking for work) compared to 81 percent for all urban U.S. men and 66 percent for poor urban men. That urban dropouts nationally experienced a higher probability of labor force activity (61 percent) than Chicago high school graduates (58 percent) suggests that differences in the availability of jobs are partly responsible for the pervasive joblessness in the inner city. Put more forcefully, the multivariate results indicate not only that urban high school dropouts confront better employment opportunities than Chicago parents who failed to complete secondary school, but that they also confront better job prospects than high school graduates residing in Chicago's inner city.

Women average lower participation rates than men, yet educational differentials in market activity by place are similar to those observed for men. Specifically, the highest probability of labor force participation corresponds to urban high school graduates nationally and the lowest corresponds to high school dropouts residing in Chicago. Among Chicago's female high school dropouts, the probability of labor force participation in a given year was only half that of all urban women—15 and 30 percent, respectively. Poor urban mothers were more likely to participate in the labor force in a given year than mothers residing in Chicago, which also highlights the importance of job opportunities in determining labor market activity.

Finally, sex differences in labor force participation are more pronounced in Chicago's poor neighborhoods compared to other urban settings. In Chicago's inner-city neighborhoods the probability of labor force participation was 2.2 times higher for male graduates compared to female graduates and 3.1 times higher for male dropouts compared to female dropouts. For all urban parents, the comparable sex ratios were 1.6 for graduates and 2.1 for dropouts, and for the urban poor graduates and dropouts, the sex ratios were 1.9 and 2.3, respectively. Most likely this reflects the higher shares of women on public assistance among Chicago residents compared to urban dwellers nationally, including the urban poor.

Acquiring labor force experience significantly increases the odds of working in a subsequent year, and as figure 6.2 shows, the employment returns to experience are appreciable.[15] The lowest probability of labor force activity in any given year corresponds to men and women who have no prior work experience, especially those who reside in Chicago's inner city. This suggests that the character of joblessness differs in Chicago and

15. Our statistical model controls for age, so that experience effects do not capture the differing age composition of the respondents.

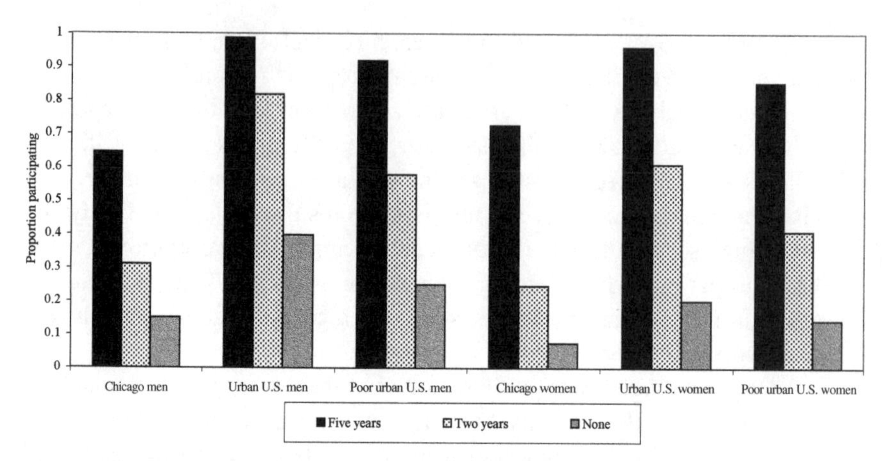

Figure 6.2 Experience Effects on Labor Force Participation: Chicago and Urban U.S. Men and Women

central cities nationally, inasmuch as married women's joblessness often is voluntary and possibly because spouses' earnings permit an intrahousehold division between market and nonmarket responsibilities. Large shares of inner-city mothers lack spouses, as shown in chapters 3 and 4; hence, their joblessness implies high reliance on means-tested income, as discussed in chapter 5. Similar constraints seldom apply to men's labor force behavior; therefore, Chicago fathers lacking any work experience are twice as likely as Chicago mothers with no experience to work in any given year.

Men and women with five years of work experience have the highest probability of labor force activity in any given year, particularly outside of Chicago, where the greater availability of jobs increases employment returns to prior work experience. Specifically, the probability that urban men with five years of work experience would work in any given year was 99 percent (92 percent for urban poor men), but only 65 percent for Chicago fathers. For women with five years of labor force experience, the odds of labor force participation in any given year were 96 percent for all central-city parents (85 percent for urban poor women), but only 73 percent for Chicago mothers. Again, these differences in the probability of working by place underscore the more limited job opportunities in Chicago compared to urban areas nationally during the 1980s.

Despite the appreciable differences in labor force activity along racial, ethnic, and gender lines reported in table 6.1, the multivariate analyses reveal small differences in participation among black, white, Mexican, and Puerto Rican men and women with similar levels of human capital and comparable family statuses. Figure 6.3 shows that most of the racial and ethnic differences in labor force participation reflect group differences in characteristics that are systematically related to labor force activity. That is, once group differences in human capital (both education and experience) and other correlates of labor market activity are systematically controlled, racial and ethnic differences in labor force participation are greatly diminished, and in Chicago they are essentially eliminated. However, sizeable differences persist in the odds of market activity by sex and place of residence. That is, consistent with findings for educational and experience differentials, the lowest annual probability of labor force participation corresponds to Chicago parents—about 50 percent for men and 20 percent for women. These adjusted probabilities compare with probabilities in the 70 to 80 percent range for urban men nationally and 38 to 49 percent for urban women. In fact, racial and ethnic differentials are far more pronounced among urban parents nationally than among Chicago inner-city parents.

Among men, the most visible differences in labor force participation obtain among the urban poor. Specifically, the predicted probability of labor force participation of poor urban men was 48 percent for Puerto Ricans and 65 percent for Mexicans with similar levels of human capital and family status. White and black men fell between these extremes. Racial and ethnic differences were less pronounced among all urban men, ranging from 80 percent for Mexicans to 72 percent for Puerto Ricans. The consistent pattern across samples is the high participation rate of Mexican-origin men and the low labor market activity of Puerto Rican men. Tienda

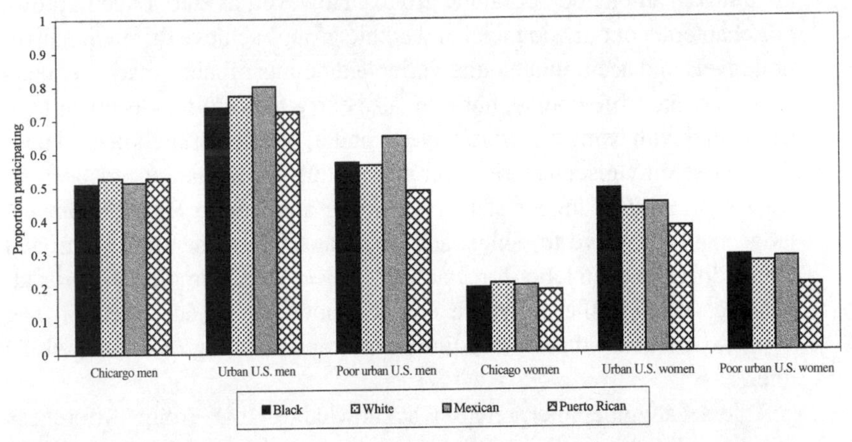

Figure 6.3 Racial and Ethnic Differences in Labor Force Participation: Chicago and Urban U.S. Men and Women

(1989) argued that contemporary differences between Mexican and Puerto Rican workers can be traced to historical differences in their preferential status. That is, while Mexicans were "preferred" workers for selected occupations that became ethnically typed, Puerto Ricans as "quasi-immigrant" citizens have never been preferred workers (see also Tienda and Wilson 1991; Hodge 1973). Hence, they benefited neither from the advantages enjoyed by immigrant and refugee resettlement programs nor from the group-specific demand for particular types of jobs.

Racial and ethnic differences in women's predicted probability of labor force participation were smaller than men's, but the range among all urban women is sizable, with a predicted probability of 38 percent for Puerto Ricans and nearly 50 percent for black women. Similarly, among urban poor women, the lowest predicted probability corresponds to Puerto Rican women, and the highest to blacks, although the absolute level of market activity is considerably lower. Nationally, urban black women experience the highest odds of labor force participation, but this is not so for Chicago's inner-city black mothers. This pattern of high black female participation rates differs from the pattern for men, but is replicated by other national data.

Although these results indicate that the marginal effects on labor force participation of racial and ethnic group membership are much smaller than the average effects, both are important for several reasons. First, the marginal effects are not zero, especially among all urban dwellers and the urban poor. Second, marginal effects assume that other characteristics that influence labor force activity, including education and work experience, are uniform among demographic groups. However, as elaborated in previous chapters, not only do racial and ethnic groups achieve different education levels and accumulate quite variable amounts of labor market experience over their life course, but they also carry forward the disadvantages associated with being reared in poverty and in parent-absent homes. These life-course circumstances reverberate on adult labor market prospects in ways that exclude higher shares of blacks and Puerto Ricans from the labor market relative to whites and Mexicans. Thus, the large racial and ethnic differences in labor force activity reflect the accumulation of disadvantages through the early life course—notably limited education and work experience—that in turn limit opportunities along racial and ethnic lines.

A less sanguine interpretation acknowledges that group differences in accumulated work experience represent minority workers' legacy of unequal treatment in the labor market and pre-market discrimination. If minority workers, especially black men, acquire less experience because of difficulties in locating jobs and receiving job offers, labor market experience captures the accumulated consequences of prior market discrimination. This explanation focuses not on the marginal employment returns to labor market experience, but on the lower average stocks of work experiences that minority workers bring to the market in any given year. Statistical tests to evaluate whether the returns to education and work experience differ for racial and ethnic groups confirmed uniform effects for minority and nonminority parents. Such evidence is generally interpreted to indicate that discrimination does not produce unequal access to jobs among black, white, Mexican, and Puerto Rican parents. However, it would be premature to conclude that discrimination does not operate to limit labor market opportunities for minority workers, because pre-market discrimination and disadvantages experienced early in the life course effectively limit inner-city residents' labor market options as adults.

A second major finding is that labor market opportunities figure prominently in determining the odds of labor force activity. That the lowest participation rates correspond to Chicago parents is consistent with the story of industrial restructuring and the spatial mismatch hypothesis. But it is also possible that the differences in work activity across samples partly reflect differences in willingness to work—an idea we investigate below. Yet, even taking into account a host of individual characteristics that influence market behavior, our results consistently show lower participation levels in Chicago compared to urban parents nationally as well as poor urban parents. Although this finding does not prove that structural factors—most notably, limited job opportunities—are responsible for lower levels of work activity in Chicago's inner city, it poses a testable proposition that we scrutinize more closely using the UPFLS survey data. Because economic well-being is influenced not only by access to jobs but also the continuity of employment, we examine racial and ethnic differences in job separations. This inquiry addresses allegations that minority socioeconomic disadvantages derive from their disproportionate concentration in marginal jobs.

Labor Force Exits and Reentry

The likelihood of being employed in a given year, which is the basis of racial and ethnic differentials in labor force participation and joblessness evaluated above, depends on the probability of remaining employed upon securing a job. If minority workers are more vulnerable to job losses, either because they have access to the more precarious jobs or because their minority-group status renders them more vulnerable to a separation than comparably skilled white workers (Hodge 1973), then it is important first to establish whether minority workers do in fact lose jobs at a higher rate than nonminority workers. The second task is to ascertain whether and how much individual and structural factors govern reentry to the labor force. Although structural factors have been invoked to explain pervasive joblessness among inner-city residents, there is surprisingly little evidence demonstrating that minority workers, especially blacks, are more vulnerable to joblessness because they are more likely to work in less secure jobs.

Analyses of job exits and labor force reentry following a jobless episode are based on life table methods described in Appendix B. Entry and exit dates for each job permitted a fairly accurate measurement of the duration of employment episodes for both the national and Chicago surveys. The survivor functions reported in figures 6.4 and 6.5 show the probability of remaining in a job for each year since respondents began a specific employment episode for men and women, respectively. Job reentry measures the duration of joblessness starting at the date of job termination and culminating when a new job begins. The corresponding survival functions shown in figures 6.6 to 6.8 indicate the probability of remaining jobless for each year since respondents exited the last work episode.

The higher levels of joblessness and employment instability among black and Puerto Rican inner-city parents reported in tables 6.1, 6.4, and 6.5, respectively, imply higher rates of job separation relative to white parents. Figures 6.4 and 6.5 display the survivor functions representing the proportions of black, white, Mexican, and Puerto Rican men and women living in Chicago and U.S. central cities, respectively, who remained employed at successive durations measured in years.

Figure 6.4 reveals much less job stability among inner-city men compared to urban men nationally. Not only are the survivor functions steeper in Chicago compared to those for urban parents nationally, indicating appreciably less persistence in jobs, but the patterning of racial and ethnic differences varies across samples. Specifically, after one year on the job, over 92 percent of urban men remained employed (only 89 percent of urban poor men) compared to less than 82 percent of Chicago men. After five years, 71 percent of urban fathers remained in the same job, but only between 30 percent (black, white, and Puerto Rican men) and 40 percent (Mexican men) of Chicago men did so. And whereas over 60 percent of urban men remained in the same job for ten consecutive years, only about

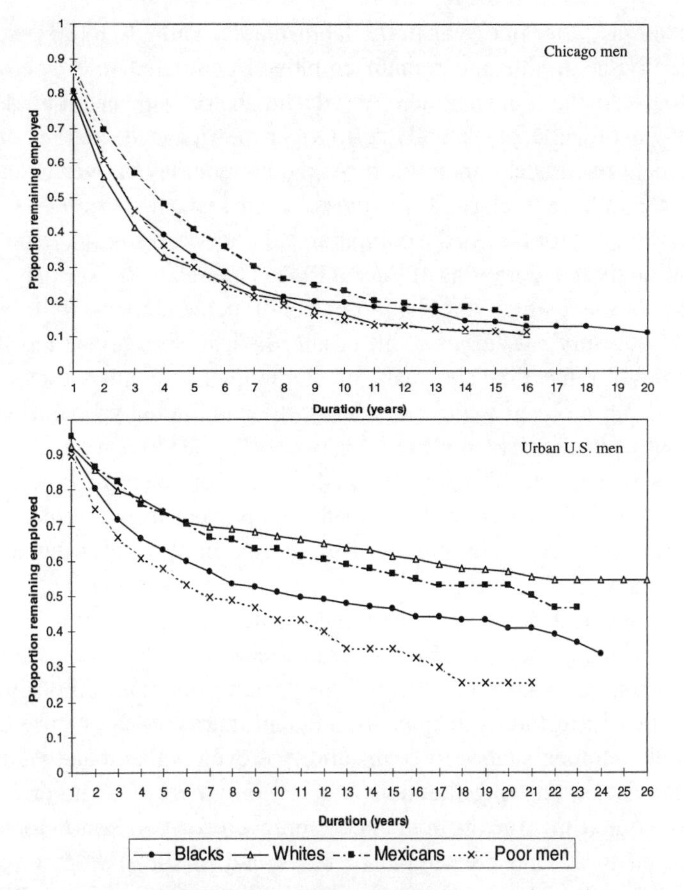

Figure 6.4 Work Exit Rates: Chicago and Urban U.S. Men

one in five Chicago men experienced decade-long continuous employment. These results, which are highly consistent with descriptive evidence showing a high level of employment instability throughout the life course among Chicago residents (tables 6.4 and 6.5), do not indicate whether job turnover results because of voluntary reasons (higher quit or dismissal rates) or involuntary reasons (jobs are eliminated or scaled down).

Racial and ethnic differences in the duration of jobs reveal more differentiation nationally compared to Chicago. That is, not only are the national survival curves flatter, but they fan out over longer durations. Among all urban men, racial and ethnic differences in the pace of labor market exits emerge after two years on the job. In addition to leaving work at a higher rate than urban fathers nationally, Chicago men exhibit differential rates of labor market exit before two years of employment.

For example, after one year in the labor market, only 79 to 80 percent of whites living in Chicago remain employed compared to 87 percent of Puerto Rican and Mexican men. And during the second year at work, 40 to 44 percent of white, black, and Puerto Rican men leave the market, but only 30 percent of Mexican men do so. At the national level, over twenty-one years would have to elapse for whites to achieve a comparable exit rate of 44 percent, but for blacks the comparable duration is only six years.

Nationally, job durations of Puerto Ricans are relatively similar to those of all poor men, which imply exit rates of 40 percent after only four years on the job. Survival curves of all urban Mexican men reveal the slowest exit rates for the first six years of employment but thereafter diverge slowly from those of white men. Thus, the conditional probability of remaining continuously employed for twenty consecutive years is 57 percent for whites and 53 percent for Mexicans. In Chicago, however, the steeper survival curves imply that only 16 percent of white men and 23 percent of Mexican men remain continuously employed for ten consecutive years.

For women, there is less racial and ethnic differentiation in work exit rates in Chicago compared to all central cities, and the survival curves are steeper (see figure 6.5). In this respect, women are similar to men. Owing partly to family responsibilities and partly to the nature of jobs they hold, women's labor force instability is even higher than that of men. Conditional on entering the labor market, only half of white and Puerto Rican women in Chicago remained employed for two continuous years compared to 52 and 58 percent of black and Mexican women, respectively. The high labor market exit rates imply that 16 percent of Mexican and 13 percent of black women remained employed for ten continuous years, but only about 10 percent of white and Puerto Rican Chicago mothers experienced job spells that lasted a decade or more.

Nationally, the highest labor market exit rates correspond to all poor women, and they most closely resemble the survival curves of Chicago's Mexican-origin women both in shape and level. Yet, for most groups, the job exit rates are approximately twice as high in Chicago compared to all central cities. For example, black mothers residing in Chicago's inner city experienced a 24 percent probability of remaining continuously employed for five consecutive years compared to 49 percent of their national urban counterparts. For whites, the comparable shares were 21 percent and 46 percent, respectively. Black urban women's survival curves show slower rates of exit from the workforce, which run counter to images of low market attachment among them. Finally, the similarity of urban poor

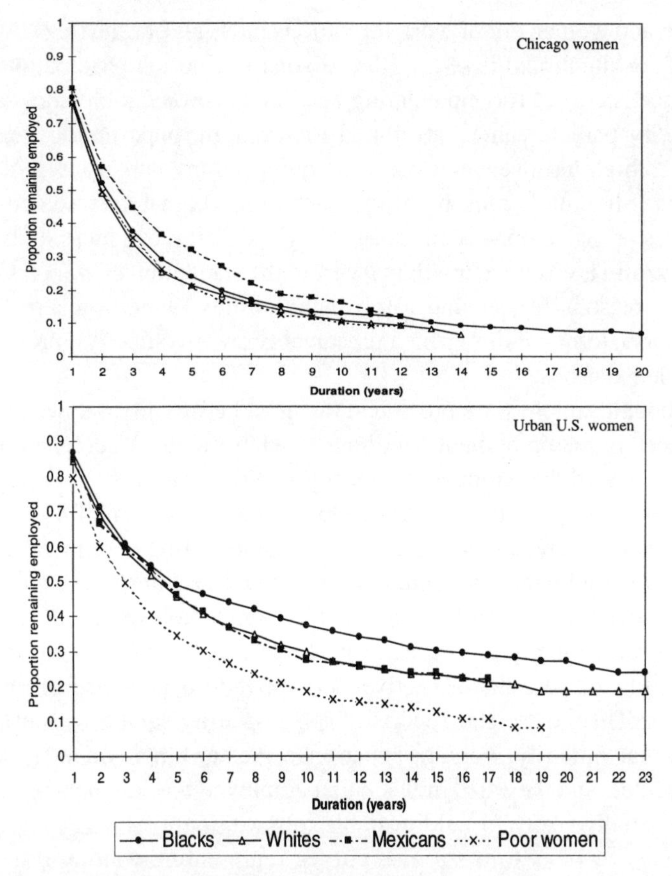

Figure 6.5 Work Exit Rates: Chicago and Urban U.S. Women

mothers' survival curves to those of inner-city mothers implies that common circumstances prevent both inner-city and all poor mothers from remaining in the labor force. These circumstances are graphically described by Edin and Lein (1997) in *Making Ends Meet* and by respondents to the Social Opportunity Survey.

Of course, not all labor market exits imply chronic joblessness, even in slack labor markets. Some workers will remain out of the market for long periods, while others will reengage almost immediately. Hence, the exit rates reported in figures 6.4 and 6.5 have very different implications for economic well-being. To further consider what factors influence labor force careers of inner-city parents, we examined the waiting times to reemployment among those who experienced a job separation. Using life table techniques (described in appendix B), we estimated the proportions of men and women out of work for various intervals of time (e.g., "waiting times," or durations) based on the reasons for the job termination. Such evidence is crucial for appreciating how much chronic joblessness among inner-city parents can be attributed to shrinking opportunities and how much to high turnover rates based on quits and involuntary terminations. Among "structural" reasons, we included reports that a firm went out of business or that a job was eliminated (e.g., labor force reduction), because these terminations were involuntary from the standpoint of workers. "Individual" reasons for leaving a job include quits (which presumably are voluntary), terminations (which presumably are involuntary), promotions, and other reasons.[16]

Although sample sizes precluded racial and ethnic breakdowns in waiting times to reemployment, auxiliary tabulations about perceived job security revealed that employed minority workers felt their jobs were quite precarious (see Tienda and Stier 1996a, table 3). About half of employed black men perceived themselves at high risk of losing their job to a plant-closing or employment reductions, compared to one-third of Hispanic men and just over one-fourth of white men. Among employed mothers, blacks were twice as likely as whites to perceive that their jobs were insecure (40 versus 20 percent, respectively) compared to one-third of Hispanic mothers. Thus, these self-reports of job insecurity square with prior evidence that minority workers experience greater employment instability than whites, and they also indicate that employed black, inner-city parents are especially likely to hold unstable jobs.

Figure 6.6 plots four survival curves representing waiting times until reemployment for jobless inner-city mothers and fathers according to individual or structural reasons for their job exit. The horizontal axis represents years since a job ended until a reemployment occurs, and the vertical axis shows the proportions who remain jobless. Protracted joblessness is indicated by the rightmost tail of the curves. Comparable analyses were not possible for the national sample because information needed to distinguish exits stemming from structural versus individual factors were not available.

These survival curves yield several insights about reemployment probabilities. First, protracted joblessness is more prevalent among women than among men, as evident in the flatter survival curves and their higher placement vis-à-vis men's curves. Joblessness is probably not equally voluntary

16. Our results are similar whether promotions are excluded or included among individual factors. This is because, of 5,100 spells analyzed, only 45 (<.01 percent) involved promotions.

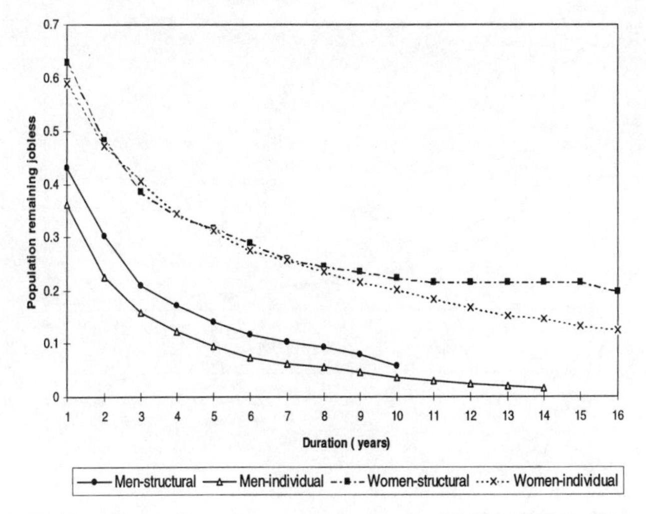

Figure 6.6 Waiting Time to Reemployment by Reason for Job Exit: Jobless Chicago Men and Women

for men and women. This is because women's exits are more often related to family responsibilities, and because caretaking as sole parents effectively limits their job options. Also, job terminations precipitated by economic dislocations (i.e., plant closings or workforce reductions) are more consequential for men than for women. This is evident in the higher placement at all durations of the "structural reason" curve relative to the "individual reason" curve for men. For instance, one year following a job separation, 43 percent of inner-city men whose job exits occurred because of establishment closings or workforce reductions remained without work compared to 36 percent of men whose exits resulted from quits, terminations, or other individual reasons. After three years, 21 percent of men whose jobs disappeared when plants closed or workforces were streamlined remained out of work compared to 16 percent of men whose jobs ended because of firings, quits, or other reasons. Five percent of men whose employment terminated because of plant shutdowns or other forms of restructuring remained jobless for ten consecutive years, but among those who terminated employment for individual reasons, less than 4 percent remained jobless this long.

Compared to men, joblessness for women was less sensitive to reasons for exiting. For instance, two years following a job exit, approximately half of all women remained out of the labor force, and there were no differences according to reason for leaving. Approximately one-third of inner-city mothers did not reenter the labor force four years after leaving,

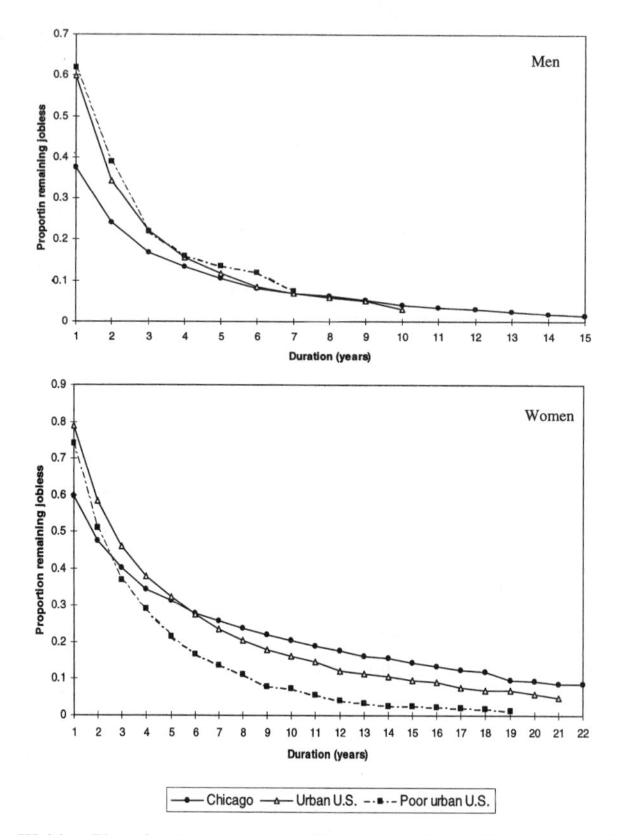

Figure 6.7 Waiting Time for Reemployment: Chicago and Urban U.S. Men and Women

and about 20 percent remained out of the labor force for ten consecutive years, irrespective of the circumstances that precipitated their exit. Beyond ten years of joblessness, structural reasons appear to be more influential than individual reasons in prolonging the waiting time to reemployment, but estimates of differences are less stable because the number of observations is too small to discern significant deviations.

To examine whether some workers faced higher risks of job loss due to the economic dislocations occurring in Chicago, we examined sex as well as racial and ethnic differences in reemployment probabilities, which are reported in figures 6.7 and 6.8. Figure 6.7 compares men and women, respectively, in Chicago and all urban areas as a way of addressing Wilson's (1996, 1987) thesis that chronic inner-city joblessness was due to the structural changes that drastically reduced job opportunities for inner-city residents. Figure 6.8 examines racial and ethnic differences *within* the inner city to determine whether blacks were disproportionately affected by the structural changes governing the availability of jobs in Chicago.[17]

As in the previous graph, protracted joblessness is indicated by the right tail of the curve, and the steepness of the curve reflects the pace of labor force reengagement: steeper curves indicate a faster pace of reemployment because the proportion remaining jobless is lower. Thus, the top panel of figure 6.7 indicates that, conditional on having held and lost a job, 60 percent of urban men, including the urban poor, remained jobless for one full year compared to only 38 percent of Chicago's inner-city men. Although this would appear counterintuitive at first, it is important to realize that employment rates are much lower in Chicago than in central cities nationally, and that, conditional on securing employment, job turnover is much higher in Chicago. After five years following a job loss, the proportion of men residing in Chicago and all central cities who remain without work is approximately equal: around 11 to 12 percent.

Both Chicago and the national urban sample include a segment of the male population that is difficult to employ. Seven years following a job termination, about 7 percent of men remained jobless; this share of men experiencing protracted joblessness is uniform across settings and includes urban poor men. Although chronic joblessness is a serious policy problem, and chronic unemployment for five years or more can have devastating personal and family consequences, our estimates indicate that jobless episodes exceeding ten or more years are typically rare and involve a very small share of the male population.

For women, reemployment rates were slower than for men, and the length of jobless spells was appreciably longer, particularly among Chicago inner-city residents. In other words, the pace of labor market reentry was faster for all urban women, and even more so for the urban poor, compared to the pace for Chicago inner-city women. One year following a labor market exit, 75 to 80 percent of urban women remained jobless, indicating that only 20 to 25 percent of job leavers had jobless spells of a year or less. Approximately 40 percent of Chicago women reentered the labor force within a year of a job separation. Five years after exiting the labor market, the reentry rates of Chicago women converged with those of all urban women, by which time nearly two-thirds were

17. Because of sample sizes, it was not possible to disaggregate by race, ethnicity, and sex for both the national and Chicago populations. Therefore, we compare men and women using the Chicago with the national samples, and for Chicago, we also examine racial and ethnic differences for men and women.

reemployed. Thereafter, reemployment rates of all urban women surpass those of all Chicago women. Conditional on having held and left a job, 20 percent of Chicago women remained jobless in excess of ten years compared to 16 percent of all urban women. That the duration of jobless spells was considerably shorter for poor urban women than for all urban women may be due to welfare-to-work programs targeted at low-income women.

In Chicago, differences in reemployment rates conditional on having held a job are not uniform among black, white, Mexican, and Puerto Rican parents. Two main differences between the reentry profiles of men and women are the flatter curves of women, indicating slower rates of labor market reentry, and the length of the curves, reflecting longer jobless spells for women compared to men. These differentials are reported in the two panels of figure 6.8, which address whether minority groups have a more difficult time securing jobs in Chicago.

Consistent with evidence reported above, black men and Puerto Rican women experience the greatest difficulties reengaging in the labor market even after having held a job. This is evident in the higher placement of the black survivor curve above that of white, Mexican, and Puerto Rican men in the upper panel of figure 6.8 and the higher placement of the Puerto Rican survivor curve above that of black, white, and Mexican women in the lower panel. Black men also average longer jobless spells than their Hispanic and white counterparts, as over 5 percent remained without work twelve years after leaving a job. However, no other group of inner-city fathers remained jobless for a dozen consecutive years, and only 3 percent of white men were without work for a decade following a job loss. Among women, approximately 8 percent of black and white mothers were jobless for nineteen consecutive years following a job separation, but twice as many Puerto Rican women were chronically unemployed that long, which is consistent with the evidence reported in chapter 5 showing that the longest welfare spells correspond to Puerto Rican mothers.

On balance, but especially for men, results reported in figures 6.6, 6.7, and 6.8 indicate that inner-city residents experience greater difficulties keeping their jobs and securing new jobs if they lose them. However, these findings do not address *why* minority workers appear to have greater difficulty obtaining jobs, and in particular why black reemployment rates are lower than those of Mexicans, who have far less education. The main contending explanations are that minority inner-city parents, and blacks in particular, are unwilling to work in jobs that offer lower wages and less attractive working conditions relative to those available during the

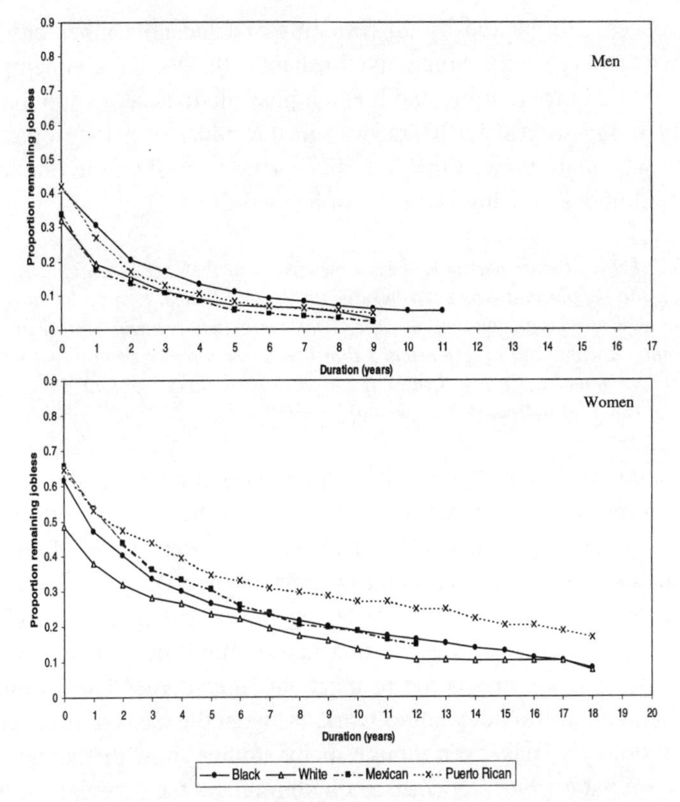

Figure 6.8 Waiting Time to Reemployment: Chicago Men and Women by Race and Ethnicity

1950s and 1960s, and that discrimination remains a powerful force in maintaining racial and ethnic labor market inequality. We consider these possibilities next.

Reservation Wages and Willingness to Work

> *Look, however difficult the job is, even if it pays minimum wage, it's a good job, because you are getting something that you didn't have before. Now you can pay your rent and bills and food and so on, it's a good job. However bad, any job is a good job.*

Ramiro, a Mexican father of three who has been working thirteen years as a casting machine refitter in the same firm, sees work as a necessary aspect of life. In his view, any job is better than no job; hence, working conditions should not determine whether or not to accept a job. This view is not necessarily shared by all residents of Chicago's inner city, even those who want to work. Some, like Lashanda, thirty-four, an unemployed black mother of four, argue that it is not possible to accept a job that does not pay living wages. Lashanda was in the midst of a long episode of welfare when interviewed in 1987 (since 1980) and living in one of Chicago's infamous housing projects. In her words,

> *Well, I think the immigrants, they come over and they get jobs quicker than the people that was born here in the United States, I guess because they willing to work minimum wage. The immigrants are willing to work minimum wage and the Americans that live here, some of them might will work minimum wages. So I think it's kind of tipsy-tursy like that. Because some [people] will work and [some] people won't.*

These contrasting statements by a Mexican immigrant and a black native-born resident instantiate the popular claim that immigrants compete with low-wage domestic workers and undercut their wages. Like many respondents interviewed, particularly African Americans, the inner-city poor perceive that immigrants are taking jobs that they might rightfully claim, but an alternative view is that native minorities refuse low-wage jobs, preferring welfare as the primary income source. The assumption that jobless adults do not want to work has been the basis of great controversy in policy arenas, even though many studies show this allegation to be false (Wilson 1996). Of course, single mothers face special hardships, because childcare costs usually exceed the incomes provided by low-wage jobs, which seldom provide health insurance or other crucial benefits (Edin and Lein 1997). Yet, almost without exception or qualification, responses to a question about preferences for work over welfare reported in chapter 5 conform with the view that employment is preferred to living on public aid.

That expressed preferences often do not conform with actual behavior warrants a more systematic scrutiny of parents' willingness to work. What do respondents mean when they say they want a job? In particular, will they accept a minimum-wage job if offered one? More generally, under what conditions will inner-city parents take a job? And—relevant to our interest in why racial and ethnic differences in labor force activity persist—do immigrants take jobs away from native blacks by accepting jobs that pay less than the minimum wage?

To address these questions about willingness to work, we restrict our analyses to Chicago because the National Survey of Families and Households lacks relevant information for a parallel analysis. However, the employment section of the Chicago survey was tailored to address questions about the minimum conditions that parents require in order to enter the labor force. All respondents who were jobless (either unemployed or out of the labor force) when interviewed were asked whether they wanted a job now and, if so, the number of days they would be willing to work per week. They were also queried about the salary and wages they required to accept a job.[18] These items are used to determine what labor economists call "reservation wages," which refer to the minimum standards required by prospective workers in order to accept a job offer.

Contrary to images perpetuated by popular writings about the urban underclass, Tienda and Stier (1996b) report that around 90 percent of jobless black fathers and 100 percent of jobless Mexican fathers said they wanted a regular job and were willing to work at least five days per week.[19] Yet only 82 percent of Puerto Rican fathers and about three out of four white jobless fathers reported they wanted a job. In other words, among the jobless, it was white men rather than minority men who were least likely to report a "willingness" to work. Given the notoriously high rates of labor force activity among Mexican immigrants and the disproportionate representation of the foreign-born among Mexican-origin fathers included in the survey, their unanimous willingness to work is hardly astonishing, especially since virtually all had worked in the past (see table 6.1). Because 17 percent of black men were not looking for work (and half of these had never worked), their virtually unanimous desire for a job is a bit more surprising, and perhaps even suspect. At a minimum, this disparity between actual behavior and reported preference warrants further scrutiny, all the more so because three out of five jobless men reported public aid as their main means of support at the time of the survey, while less than one-third had access to unemployment benefits.

Among women, racial and ethnic differences in reported willingness to work were even sharper than for men. Nine out of ten black women who were jobless when interviewed in 1987 indicated they wanted a regular job, compared to three out of four Hispanic (both Mexican and Puerto Rican) mothers. Surprisingly, white jobless women were least disposed to enter the labor force, and many of those who claimed they wanted a

18. The exact wording of the question follows: 6. A. What would the wage or salary have to be for you to be willing to take a job? If R responds "Minimum Wage," re-ask A. Probe if necessary: Is that per hour, day, week, or what?

19. The small sample sizes of jobless white and Mexican fathers require caution in interpreting results.

regular job preferred less than full-time work. Also, black jobless mothers who were willing to work indicated they wanted to work at least five days per week. Over three-fourths of jobless black mothers were supported by public aid, while only 20 percent received unemployment benefits. By contrast, only one-fourth of jobless Mexican-origin women relied on public aid as a means of support, while over 70 percent were supported through labor income produced by their husbands (See Tienda and Stier 1996b).

Comparisons between actual wages received by employed parents and reservation wages aspired to by jobless parents, which are reported in table 6.6, reveal striking racial and ethnic disparities, as well as sex differences. Average wage rates of employed men exceed those of employed women by 20 percent (blacks and whites) to 50 percent (Mexicans). Also, a clear

Table 6.6 Average Wages and Means of Support: Chicago Men and Women by Race, Ethnicity, and Current Employment Status (Means and Percentages)

	Black	White	Mexican	Puerto Rican
Employed Men				
Wage rate	$8.90	$14.60	$7.90	$8.00
[*N*]	[225]	[111]	[213]	[118]
Jobless Men				
Reservation wage	$5.80	$9.30	$7.20	$6.20
Means of support				
Aid	60.8	33.6	33.8	44
Means of Support				
Family	5.4	—	13.5	13.2
Work benefits	27.3	46.6	33.8	16
Informal employment	6.4	19.8	19	6.9
[*N*]	[83]	[16]	[15]	[30]
Employed Women				
Wage rate	$7.30	$12.00	$5.20	$6.10
[*N*]	[314]	[125]	[133]	[103]
Jobless Women				
Reservation wage	$5.50	$6.60	$5.30	$5.30
Means of support				
Aid	77.5	41.4	24.4	68.8
Means of Support				
Family	1	4.4	2.4	0.4
Work benefits	20.3	52.7	70.9	29
Informal employment	1.2	1.5	2.3	1.9
[*N*]	[45]	[112]	[128]	[203]

Source: UPFLS

wage hierarchy exists among employed parents, irrespective of sex, with whites receiving the highest wages, Hispanics receiving the lowest wages, and blacks between these extremes. Racial and ethnic differences in wage rates among inner-city men are somewhat surprising, because most studies show that Hispanics out-earn blacks. One possible explanation is that employed inner-city black parents may be a more selective segment of their source population because a relatively smaller share of them were employed at the time of the survey (see table 6.1). Differential selectivity of employed versus jobless parents has implications for comparisons of actual and reported wages, which we address subsequently.

That the reservation wages of jobless workers are systematically lower than the actual wages of employed workers undermines arguments about unrealistic wage expectations as a major reason for the pervasive joblessness of inner-city residents and minorities in particular. Among inner-city residents, whites—not minorities—reported the highest reservation wages. Moreover, among men, blacks rather than Mexicans expected the lowest pay in order to accept a job, despite their higher educational credentials. Reservation wages of Mexican men are most consistent with the actual wages of their employed counterparts, while the discrepancy between the average wages of employed black men and the reservation wages of their jobless counterparts was largest among the groups compared. Compared to jobless men, reservation wages of jobless women exhibit less differentiation, except that whites report significantly higher wage expectations than their minority counterparts.

Although the evidence reported in table 6.6 challenges common allegations that inner-city joblessness results from unrealistic wage expectations of minority workers, in fact these tabulations cannot answer whether the wage demands of jobless parents are realistic relative to prevailing wage opportunities and their skill endowments. This is because employed and jobless parents may differ in ways that are systematically correlated with wage offers, as suggested above for black men. In other words, the subset of employed parents may have higher education, more work experience, or live in better neighborhoods compared to jobless parents. Therefore, it is incorrect to compare directly the *reservation* wages of jobless individuals with the *actual* wages of those employed.

To address this problem, we first computed an index that captured the selection regime representing how employed parents differ from jobless parents in several characteristics that influence the likelihood of labor force participation. Appendix table B6.5 reports these statistical results. Subsequently, we computed a standard wage function for employed men

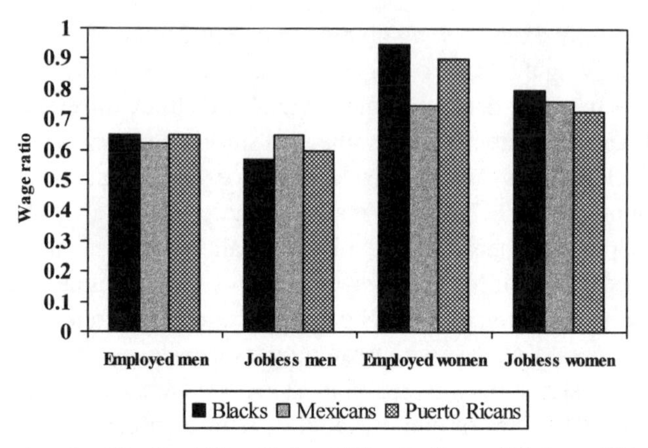

Figure 6.9 Actual and Predicted Wage Ratios of Minority Groups Relative to Whites, by Employment Status and Sex

and women that takes into account the selection regime governing their participation decision (reported in appendix table B6.6). From this model, we generated a set of expected wages for both the jobless and the employed.[20] Figure 6.9 summarizes these results.

Racial and ethnic differentials in predicted wages tell a very important story about the significance of color and opportunity in Chicago, because the predicted wages are based on a standardization that makes groups equivalent in terms of their characteristics, especially human capital endowments, and the focus on a single labor market presumes that comparably skilled workers have access to relatively comparable opportunities.[21]

The predicted wages depicted in figure 6.9 indicate the approximate offers that jobless respondents can expect in the Chicago market, given where they live, their education, prior work experience, and their minority-group status. In general, predicted wages are lower than actual wages received by employed parents or the reservation wages of jobless parents, but the deviations between predicted and actual/reservation wages differ between men and women, and also among blacks, Mexicans, whites, and Puerto Ricans. Several types of comparisons are possible, but our interest in the "color of opportunity" makes the disparities between white and

20. We discuss the technical procedures to derive these results and the statistical results used to generate the predicted wages in appendix B. These results are reported in appendix table B6.7.

21. The first-stage equation to predict the labor force participation decision includes neighborhood poverty status as a predictor. Inclusion of neighborhood poverty status is appropriate for this analysis because the focus is on current employment status and because the goal was to model as many covariates as were theoretically and empirically relevant for the participation decision.

minority men and women and the jobless gap between expected and predicted wages most interesting.

For employed men, the ratio of minority to white wages hovers around .64, which means that men of color are compensated only two-thirds as much as comparably skilled whites. Among employed women and also the jobless, predicted wages of blacks were virtually identical to those of whites, but Mexican-origin women earned only three-fourths of what comparably endowed white mothers received. Puerto Rican women were somewhere between these extremes: their wages were 90 percent of those of comparably skilled whites. Thus, among employed parents, especially fathers, our results demonstrate that *in Chicago's labor market, color matters.* These results are consistent with a discrimination story.[22]

Parallel comparisons of predicted wages between jobless minority and nonminority inner-city residents reinforce this conclusion, which underscores the persisting disadvantages associated with minority-group status in Chicago's labor market. Among jobless men, Mexicans and Puerto Ricans could expect wages about 60 to 65 percent of those received by white fathers with similar characteristics, but black jobless fathers could command only half as much. Although jobless women could expect much lower wages than jobless or employed men, minority women would earn only about 80 percent as much as their comparably endowed white counterparts.

A comparison of the predicted wages for employed and jobless parents would lend credence to claims that unrealistic reservation wages are responsible for the low employment rates of inner-city workers if the predicted wages of the jobless exceed the predicted wages of their employed ethnic counterparts. Again, this conclusion finds no support for any of the groups compared. In fact, the *lowest* expected-wage ratio between jobless and employed residents corresponds to blacks, hovering around .75 for both sexes, and this is so even though blacks have the highest rates of joblessness. For whites and Mexicans, the ratio of expected wages for jobless relative to employed workers ranges from .9 to .95; Puerto Ricans fall between these extremes with ratios ranging from .75 to .81.

From the evidence presented, it is difficult to argue that the reservation wages of jobless inner-city residents are out of line with those of comparably skilled workers and therefore might be responsible for the large racial

22. We measure discrimination as a residual after modeling relevant determinants, but measurement of actual discrimination would require an audit methodology or surveys with employers. These studies also would reveal pervasive actual discrimination (Fix and Struyk 1993).

and ethnic differentials in labor market status reported in table 6.1. But for Chicago's inner-city parents, there is a clear penalty for color—that is, for being a member of a minority group, especially for being black. Despite the role of unions in protecting the wages of semiskilled and unskilled workers in the past, union membership is not a viable explanation for the wage advantages enjoyed by white parents, because our statistical analysis considered union membership and establishment size as predictors. Besides, white workers report *lower* rates of union membership than minority parents, yet their wages are appreciably higher. Therefore, we conclude that a statistical discrimination story is more consistent with these results than a reservation wage story, not only because of the large racial disparities in expected wages, but also because reservation wages are "hypothetical" until a job offer is actually tendered. For black and Puerto Rican men, the lack of job offers, not low wages, is the more salient aspect of limited opportunity.

In emphasizing statistical discrimination, we do not mean to discount the importance of education, which is a significant prerequisite for labor market success. But recall that black respondents were more likely to be jobless, and they reported the lowest reservation wage rates, yet they average considerably more education than Mexicans or Puerto Ricans. However, employers are increasingly reluctant to hire them, particularly when alternative sources of immigrant labor are so readily available. Put bluntly, employers seem to prefer Mexican and white workers over blacks and, to a somewhat lesser extent, Puerto Ricans. This conclusion was confirmed by numerous interviews with employers, which also revealed that signals about where workers live also discourage employers from hiring inner-city minorities (Kirschenman and Neckerman 1991).

Therefore, a simple demand story—that is, one emphasizing the decline of unskilled jobs that paid living wages—cannot adequately explain the large and persisting racial and ethnic differences in a single labor market. Over thirty years after major civil rights legislation was enacted, color—that is, minority-group status—continues to delimit economic opportunity in Chicago as well as other urban areas. It appears that the problem of the twenty-first century will remain the problem of the color line.

Conclusions

This chapter opened with evidence that inner-city parents participated in the labor force less and experienced appreciably higher levels of joblessness than urban parents nationally. Further, inner-city parents who lived in high-poverty neighborhoods experienced higher levels of joblessness than those living in less poor neighborhoods. Joblessness *is* pervasive in the inner city, and its occurrence as well as the reasons for its occurrence vary appreciably by group membership and sex. By themselves, such stylized "facts" do little to clarify the role of color and opportunity in generating and maintaining racial and ethnic labor market inequality. This is because contemporary economic inequities reflect group differences in skill endowments, which themselves are a product of prior discrimination, limited opportunities, and early life-course events that compromise future human capital investments, as well as group differences in willingness to accept available jobs. Accordingly, our empirical analyses were designed to determine what life-cycle experiences (e.g., family status, education, and prior job episodes) and what social circumstances (e.g., plant-closings, work slowdowns, relocations) were responsible for the great diversity of labor force experiences among black, white, Mexican, and Puerto Rican inner-city residents.

That the majority of inner-city parents worked prior to the survey shows some attachment to the labor market, but the connection between prior and current labor market position depends on the continuity of work experience, early investments in education, minority-group membership, and the relative availability of jobs. For women, family status also constrains market opportunities, especially if they are the sole caretakers of their children. The majority of parents who were jobless at the time of the survey reported they wanted a job, and among those who did not want a job, family, health, and school were the major reasons for not seeking work (Tienda and Stier 1991). If these were legitimate reasons for not working in the past, they are less tolerable in a world where public assistance is the main alternative to work, where work has become a requirement for receipt of means-tested benefits, and where stringent time limits have been imposed on welfare recipients.

Human capital investments also differ among minority and nonminority men. Specifically, black fathers have an educational advantage over Mexican and Puerto Rican fathers based on years of school completed, but not in another crucial form of human capital, namely, work experience. Analyses of employment separations revealed that the sharp racial and ethnic differences in current employment status operate largely through human capital, especially the accumulation of labor market experience. Of course, acquisition of experience presumes that a job is secured in the first place, which for blacks and Puerto Ricans is less likely compared to whites and Mexicans. That is, the difficulties minority workers experience in locating jobs early in their life course accumulates to profiles of disadvantage that present as experience deficits among adults, which in turn make them less attractive to prospective employers. Thus, the job trajectories of minority parents in poor Chicago neighborhoods indicate not the absence of a work ethic, but limited opportunities to secure employment coupled with persisting discrimination against blacks and Puerto Ricans. This is evident in the higher levels of joblessness experienced by inner-city parents compared to minority parents nationally (especially minority fathers), as well as the longer wait times between jobs and lower likelihood of reemployment following a plant-closing or relocation.

Another recurring theme is that the lack of jobs in Chicago was a powerful force limiting economic opportunity. Not only is this strikingly evident in the lower levels of employment between Chicago's inner-city residents compared to residents of central cities nationally, but also in the finding that high school dropouts nationally had a higher probability of securing a job than Chicago's high school graduates. Limited job opportunities also were a recurrent theme among respondents to the Social Opportunity Survey. When queried about job opportunities in Chicago, Bernardino responded as follows:

> *No, there isn't [enough good jobs]. Because the good jobs are already taken. Since there're so many people in the United States and there are not enough good jobs that are [not] already taken, these people are going to keep these jobs. They're not going to let them go for nothing. So, the ones that have no chance to work are the ones that are unemployed now.*

An unemployed Puerto Rican father of three with an unstable work history despite having a GED and some college, Bernardino understands firsthand the value of jobs that pay family wages. Having worked as a packer, a shipper, a truck driver, a carpenter, and then again as a truck driver, he appreciates the importance of steady employment—a luxury he has not enjoyed, unlike Ramiro, a Mexican immigrant to Chicago (quoted above) who was steadily employed for thirteen years. Despite an unstable employment history peppered with episodes of unemployment, Bernardino recognizes that jobs vary in their desirability and that, at least during the 1980s in Chicago, good jobs were scarce. Jobless at the time of the survey, he perceived little opportunity—"no chance to work"—because there was little turnover among the good jobs, and there were fewer of them compared to previous times. Bernardino's comments ring with discouragement, the detachment from the workforce that often results from repeated episodes of joblessness, and demoralization about the prospects of securing work, but they also resonate to Wilson's (1996) thesis about the social consequences of declining job opportunities.

Similarly, Alyson, a black, never-married female with no children who was on general assistance when interviewed, acknowledged the difficulty of securing jobs in a slack labor market. She emphasizes the importance not only of knowing where jobs are, but also having the appropriate qualifications, including experience, to secure a job. These necessary conditions were not sufficient to guarantee employment, however, because competition for jobs was more intense during the 1980s compared to earlier periods. In her words,

> *I think you really have to know where the jobs are, what the qualifications are, because we always have to compete . . . always have to beat something . . . we always have to compete, you know, and have to beat the other person, you know, so you have to know where are the jobs and what the qualifications are, and the person that have all of these qualifications, first, and the experience, that's the person that will get the job, while the other person's thinking about what he has to have.*

Both Bernardino and Alyson identify several themes that have been addressed in this chapter, namely, the shortage of jobs in Chicago, the importance of skills (qualifications) for securing employment, and increasing competition for low-wage jobs. Although neither mentions discrimination nor direct competition with immigrants, both responses are consistent with the idea that immigration increased competition for a dwindling stock of low-wage jobs and that some people have better chances of securing jobs. In the open-ended interviews, several respondents identified the willingness of immigrants to accept lower wages than natives as a reason for increased labor market difficulties of native minorities.[23] Marcelina's view is typical. When queried whether there were enough good jobs in the city she replied, "No. Because of the majority of jobs the Mexicans are taking and they are working for less pay." And when probed what should be done to get good jobs, she responded, "They are supposed to be sending the illegals back to Mexico, and half of them are still here. I don't think they are able to do that kind of job." Marcelina, twenty-six, a single mother of one, was on welfare at the time of the interview and had never held a job despite having a high school diploma.

23. Although most evidence about employer preferences for immigrant workers is indirect, a representative survey of employers by Holzer (1996) suggests that employers prefer Hispanics to blacks for many jobs in which skill and hiring requirements are modest, but also when experience and previous training are needed. Also, large shares of low-education Hispanics are immigrants who are hired at a higher rate than blacks.

Our empirical analyses find little evidence to substantiate this claim, inasmuch as the reservation wages of all jobless workers were systematically lower than the actual wages received by employed workers, especially by minority workers. Claims that Mexicans will work for lower wages also were not supported, beliefs to the contrary notwithstanding, because black fathers systematically reported the lowest reservation wages. In fact, the reservation wages of jobless Mexican fathers were more in line with the actual wages of employed Mexican fathers. Because jobless blacks reported lower reservation wages than any group, they should be able to undercut the wages of other employed minorities and whites. That joblessness was more pervasive among blacks than any other group despite their lower reservation working conditions once again invokes discrimination as an explanation for the disparate outcomes. Recent evidence that employers prefer Hispanic to black workers, even when the former are "less skilled," further corroborates this point (Holzer 1996; Kirschenman and Neckerman 1991).

In sum, the evidence marshaled in this chapter leads us to conclude that low skills are not the major reason for the highly diverse employment experiences of white and minority parents in the inner city of Chicago or other urban places. That parents from Chicago's inner-city neighborhoods did not differ from a national sample of parents with respect to educational attainment drives home the point that formal schooling is a necessary but insufficient condition to guarantee labor market success. Employers care about skills, but they also appear to exercise preferences for workers on the basis of group membership. Whether such statistical discrimination is justified on grounds of expected productivity differences or simply reflects the enactment of prejudice will remain a matter of debate, but racial and national-origin discrimination remains a plausible explanation for the pronounced racial and ethnic differences we have documented.

In emphasizing statistical discrimination, we do not mean to discount the importance of supply-side explanations of persisting racial and ethnic differences in wages. Although our analyses of labor force experiences and wages have taken into account individual differences in education and experience, unmeasured skill differences (e.g., English proficiency) probably remain among the groups compared. Ultimately, shifts in both supply and demand curves set wages; hence, the various individual and structural interpretations reviewed at the outset are germane for understanding persisting wage disparities among black, white, Mexican, and Puerto Rican inner-city workers. Also, industrial restructuring away from unskilled jobs may partly explain the lower employability of black men and Hispanic women, but neither a human capital nor a structuralist explanation of employment suffices for the large racial and ethnic differentials in employment instability among inner-city residents. Moreover, black jobless parents reported the lowest reservation wage rates, and, among those employed, black parents also received the lowest wages. However, employers are increasingly reluctant to hire them, particularly when alternative sources of immigrant labor are so readily available.

Put bluntly, Chicago employers seem to prefer Mexican and white workers over blacks and, to a somewhat lesser extent, Puerto Ricans. Not only does color matter in Chicago, and in all central cities for that matter, but it has an exchange value that fluctuates in accordance with changes in the supply and demand for labor. Therefore, we conclude that a simple demand story—that is, one emphasizing the decline of unskilled jobs that paid living wages—is inadequate to explain the large racial and ethnic differences in a single labor market. Employer preferences play themselves out in ways that produce and maintain racial and ethnic labor market inequality.

CHAPTER SEVEN

The Contours of Opportunity

At the turn of the twentieth century, Chicago had emerged as a prominent industrial center, a major distribution center, and the nation's second largest city. The ample job opportunities created by industrial growth attracted immigrants from southern and eastern Europe along with African Americans in search of better futures for themselves and their families. As the new millennium approaches, Chicago is a different city—a place where economic opportunity has been dimmed because thousands of semiskilled and unskilled factory jobs disappeared, only to be partly replaced by low-wage service jobs. It is also a city of color, where blacks and Hispanics compose the majority of the population—over 60 percent in 1990 and even higher in the year 2000. Although its industrial decline is mirrored in its fall from the second to third largest city, Chicago remains a destination for thousands of immigrants who continue to diversify its ethnic landscape. At the beginning of the nineteenth century, the majority of Chicago's immigrants were from Europe, but at its close, the primary sending areas are Mexico, Central America, Korea, and other parts of Asia.

In the 1960s, Chicago was a focal point of civil rights activities, and in the 1980s it served as the laboratory for William J. Wilson's (1987) ideas about the emergence and social consequences of concentrated urban poverty. In describing the social transformation of the inner city produced by massive job losses and the structural transformation of employment, Wilson (1979, 1987) initially downplayed the significance of race in his writings about the causes of concentrated poverty, emphasizing instead how macrostructural changes reduced opportunities for earning a living. Presumably, the chronicity and spatial concentration of urban poverty differed from the transitory and dispersed poverty in causes, consequences, and cures, but no study compared directly whether and in what ways residents of concentrated-poverty neighborhoods differed from the poor in general. Because of his emphasis on macrostructural economic changes that altered the demand for workers of various skill levels, Wilson's (1996) key policy solutions called for changes in employment policy that would create and locate jobs in areas easily accessible to poor, inner-city workers.

That chronic poverty, joblessness, and welfare participation are much more pervasive among blacks and Hispanics compared to whites also leaves open the possibility that group membership contributed to the spread of concentrated poverty and colored Chicago's class structure. As noted in chapter 2, striking continuities exist between Chicago at the beginning and end of the twentieth century. This is because racial and ethnic competition for jobs and social positions remains a defining feature of Chicago's social fabric. However, important differences exist in the way group membership circumscribed opportunities then and now. A major difference is that jobs were plentiful at the century's beginning, and population growth fueled the economic development of the city. However, from the 1950s until 1990, the city lost population, despite the resurgence of immigration during the 1970s and 1980s. Furthermore, over the last quarter of the century Chicago lost thousands of well-paid jobs under the stress of industrial restructuring. Another important difference is that there was no welfare safety net at the beginning of the century, although recent changes in welfare policy also have thinned the social safety net for the poor, and the working poor never had much support anyway. Also, if ethnic prejudice and social injustice were the dominant forces producing racial and ethnic inequality at the turn of the twentieth century, persisting segregation and intergenerational transmission of material deprivation, with its attendant consequences for educational failure, nonmarital childbearing, and the chronic joblessness and welfare dependence they engender, are the dominant contemporary mechanisms.

Our inquiry about racial and ethnic inequality sought to answer whether and how group membership circumscribes life chances, that is "the color of opportunity," and whether the inner-city poor differ from the urban poor in their family formation, welfare behavior, and labor force attachment, as alleged by several studies during the late 1980s and early 1990s. This required establishing empirically whether discernible racial and ethnic differences appear in these behaviors and, if so, whether such differences are associated with life experience in concentrated poverty neighborhoods. From the evidence presented, we conclude that, in the main, the inner-city poor do not differ from the poor in general and that the pervasiveness of joblessness, welfare dependence, and nonmarital childbearing are due to poverty and limited job opportunities, *not to deviant behavioral dispositions.* We also show that, with few exceptions, racial and ethnic differences in these behaviors were minimal among comparably skilled adults with similar family backgrounds. Table 7.1 provides a succinct summary of key findings on which this generalization is based.

To understand how differences in economic well-being arise, we consider how decisive early disadvantages and pivotal life-course events are in shaping the life-course trajectories of minority and nonminority parents. We provide evidence about the relative importance of employment skills,

Table 7.1 Summary of Racial and Ethnic Effects (Relative to Whites) on Key Outcomes

	Outcomes				
		Pathway to Family Life			
	High School Withdrawal	Marriage	Birth	Welfare Use	Labor Force Participation
Blacks					
Chicago men	None	None	+	+	None
U.S. men	None	None	+	None	–
Poor U.S. men[a]					None
Chicago women	None	None	+	+	None
U.S. women	None	–	+	None	+
Poor U.S. women	None	None	+	None	None
Mexicans					
Chicago men	+	None	None	–	None
U.S. men	+	None	None	–	+
Poor U.S. men[a]					+
Chicago women	+	+	None	–	None
U.S. women	+	None	+	+	None
Poor U.S. women	+	None	None	None	None
Puerto Ricans					
Chicago men	+	None	+	+	None
U.S. men				None	None
Poor U.S. men[a]					None
Chicago women	+	None	+	+	None
U.S women	+	None	+	None	–
Poor U.S. women	+	None	None	None	–

[a] Blanks indicate that group effects were not analyzed due to small sample sizes.

labor market opportunity, and preferences for public assistance in explaining racial and ethnic differences in economic well-being. Our focus on how group membership circumscribes social and economic opportunity throughout the life course explicitly acknowledges that, in addition to changes in labor demand associated with industrial restructuring and the social transformation of the inner city, blacks, whites, Mexicans, and Puerto Ricans differ in their experiences with early material deprivation and disadvantages accumulated over the life course. Whereas macrostructural changes governing the industrial composition of employment shape the contours of opportunity—that is, the demand for workers of varying skill levels—early life experiences with poverty and parent absence are largely responsible for the perpetuation of disadvantage intergenerationally. The latter processes color life chances as much as, if not more than, changes in the availability of jobs commensurate with varying skill levels, that is, economic opportunities. Partly, this is because minorities, especially blacks, are more likely to begin their lives in circumstances of material deprivation which are exacerbated and reinforced by life experiences in segregated, resource-poor environments. Partly, however, it arises from the persisting significance of race in delimiting educational and economic opportunity.

We illustrate the intergenerational transmission of poverty by focusing on two pivotal circumstances that limit youth's ability to accumulate the life skills and resources required for adult economic independence, namely, high school completion and pathways to family life. A pathway refers to the timing and sequencing of life-course events—marriage and births in our application—that affect future transitions and outcomes, notably work activity, welfare dependence, and poverty. Family structure, in particular the experience of being reared by a lone parent, is a key mechanism for the intergenerational transmission of poverty and a consequence of being reared poor.

For youth, poverty is associated with premature school withdrawal, early parenting, often in the absence of marriage, and unstable labor market careers. That is, poverty both produces these outcomes and is reproduced by them. These two circumstances—failure to complete high school and early, nonmarital childbearing—have economic consequences that last well into adulthood because they compromise human capital investments and because they constrain present and future labor market options. For women, early childbearing is associated with failure to complete high school and very limited labor market prospects. For both men and women, premature school withdrawal also reduces acquisition of valuable work experience through unstable employment careers that often eventuate in labor market discouragement and prolonged joblessness.

Early life experiences with poverty more than racial and ethnic group membership per se are what ultimately "color" economic opportunity for black, white, Mexican, and Puerto Rican parents residing in poor inner-city neighborhoods. That group membership is tightly coupled with childhood poverty largely explains why blacks and, to a lesser extent, Puerto Ricans are more likely to reside in Chicago's poorest neighborhoods. That is, the economic circumstances of their families of orientation, governed as they were by past discrimination and the social consequences of residential segregation, were inadequate to prepare them to compete in rapidly changing housing and labor markets. Historically blacks have experienced the most intense discrimination, which is manifested in their higher levels of residential segregation in the poorest neighborhoods (Massey and Denton 1993). Therefore, the emergence of large racial differences in family formation, welfare participation, and employment behavior does not constitute prima facie evidence of group-specific behavior or, for those who reside in the highest poverty areas, ghetto-specific behavior. Such conclusions would only be warranted if such differences persisted after taking into account the accumulation of disadvantages throughout the life course.

In the main, we find relatively little evidence that blacks, whites, Mexicans, and Puerto Ricans differ in their family formation, welfare participation, or employment behavior or, more generally, that the behavior of the inner-city poor differs from that of the urban poor in general. Chapter 4 documents higher dropout rates in Chicago compared to urban areas nationally, but these rates were similar to those of the urban poor nationally. Thus, what appear to be place-specific effects are essentially poverty effects, because the Chicago study focused on low-income neighborhoods, where poverty rates were far above average. The noteworthy group differences in high school completion rates are those between Hispanics and whites, but these obtain nationally as well as among the urban poor and hence cannot be attributed either to ghetto-specific or place-specific behavior. Similarly, racial differences in the sequencing of marriage and births are not unique to women residing in poor urban ghettos. Nationally as well as in Chicago's inner city, black women are more likely than whites or Hispanics to enter family life via births rather than marriage. Moreover, parents who experienced similar pathways to family life exhibit no discernible racial or ethnic differences in adult household income or welfare participation. Thus we infer that childhood disadvantage and the difficult family experiences it fosters are largely responsible for the observed variations in family formation patterns and economic well-being among inner-city residents. Nevertheless, as do many other authors, we find strong evidence that bearing a child out of wedlock has lasting consequences for women's economic well-being: unwed mothers are more likely to become welfare-reliant than women who marry before bearing children.[1]

Racial and ethnic differences in welfare participation were even smaller than those observed in family formation. Contrary to popular characterizations of the urban underclass that emphasize high and chronic rates of welfare use, we find that residents of high-poverty neighborhoods are about equally as likely to rely on public assistance as are all urban poor parents. However, owing to limitations of the national survey, namely, the lack of retrospective welfare histories, we were unable to compare chronicity among the urban poor and Chicago's inner-city poor. The large differences in average participation rates among black, white, Mexican, and Puerto Rican urban parents result from group differences in family structure and income deficits that make them eligible for welfare benefits. Although Chicago's black and Puerto Rican parents were more likely to receive means-tested income than similarly endowed whites or Mexicans, these differences do not portray group-specific preferences for public assistance. Rather, they reflect the sorting processes that segregated blacks into Chicago's poorest neighborhoods and the more pervasive joblessness among Chicago's black and Puerto Rican parents. We arrived at this conclusion because we found no racial and ethnic differences in welfare participation among all urban residents or even among the urban poor who were similarly eligible for public assistance. In most respects, the urban poor behave like Chicago's inner-city residents with respect to public assistance utilization, except that the latter are more residentially segregated along racial and class lines. Thus, what is unique to inner-city residents is *not* their behavioral disposition toward participation in income maintenance programs, but rather extreme residential segregation that limits access to spatially distributed economic opportunities (Massey and Denton 1993); that socially isolates minorities from mainstream behaviors (W. J. Wilson 1987); and that truncates life chances at early ages through educational underachievement and premature assumption of family responsibilities (McLanahan and Sandefur 1994).

1. Of course, our results are based on the AFDC program, which did not impose stringent time limits on the duration of aid. Whether these generalizations will obtain for the new TANF program remains to be seen.

Residential segregation is a contemporary manifestation of a legacy of unequal treatment, and it affects current and future economic prospects by restricting access to good schools, good jobs, and good role models (Massey and Denton 1993; Wilson 1996). Among youth, residential segregation by race and income also means exposure to high-risk environments, where premature school withdrawal and teenage childbearing are common; among young adults it means limited access to a dwindling number of unskilled and semiskilled jobs that pay a living wage. The confluence of these two trends—low educational attainment of minority populations relative to whites and rising demand for skilled workers—has rendered black, Puerto Rican, and Mexican workers vulnerable to unstable work careers at best and to chronic joblessness at worst because of the skill and spatial mismatches they engender. Our analyses of labor force activity show that employment levels were extremely low in Chicago compared to urban areas nationally. Nevertheless, racial and ethnic differentials in labor force activity are similar. That is, the highest labor force participation rates correspond to whites and Mexicans and the lowest rates to blacks and Puerto Ricans in Chicago, in central cities nationally, and among the urban poor. This suggests some apparently universal forces shaping the color contours of stratification throughout urban areas, not only in cities that experienced industrial decline, although the intensity of the process may well be greater there.

We demonstrate that racial and ethnic differences in current labor market standing reflect the accumulation of disadvantages from the past, both in educational underachievement and unstable work trajectories, which eventuate in unequal stocks of work experience among adults. Blacks and Puerto Ricans accumulate less work experience than Mexicans and whites, and this circumstance, combined with low educational attainment vis-à-vis whites (particularly for Hispanics), produces group differences in current labor force activity. Stated differently, with similar levels of education and work experience, racial and ethnic differences in labor force activity would be moot in Chicago.[2] We found no evidence that the employment returns to education and work experience differ among racial and ethnic groups in Chicago. Rather, group differences in stocks of human capital are ultimately responsible for unequal rates of labor market activity. Current policies cannot undo the legacies of past unequal treat-

2. The presence of some national variation most likely reflects the variable labor market conditions across settings, which were approximately constant in Chicago (a single SMSA). These were not modeled in the national data owing to lack of relevant market characteristics.

ment, and the consequences of accumulated disadvantages in skill acquisition are magnified when labor markets slacken, as they did during the 1980s, especially in large industrial cities that lost their manufacturing base.

Finally, regarding preferences for welfare over work as a reason for chronic and pervasive joblessness in the inner city, the preponderance of evidence we present indicates that, despite high levels of joblessness and elevated rates of welfare participation, most inner-city parents prefer a steady job to reliance on public assistance. A more rigorous scrutiny of self-reports based on a multivariate analysis of reservation wages refuted arguments that inner-city joblessness results from unrealistic wage expectations of inner-city residents. Rather, our analyses indicate that wage discrimination operates to stratify incomes of employed blacks, whites, Mexicans, and Puerto Ricans; minority workers who managed to secure jobs in Chicago's slack labor market systematically received lower wages than comparably skilled white workers. For blacks, the most economically disadvantaged of Chicago's minority populations, two forces conspired to relegate them to the bottom of the wage hierarchy: they were least likely to be hired, despite their higher educational attainment relative to Hispanics, and they received lower wages than comparably skilled whites. On this basis we infer that statistical discrimination did color wage inequality in Chicago during the 1980s, above and beyond declines in the sheer numbers of jobs. The vigorous economic recovery during the 1990s reaffirms our contention that blacks are at the bottom of a hiring queue (Hodge 1973) whose contours depend greatly on the shape and vitality of the economy. Blacks have been the main beneficiaries of the current economic expansion, because they experienced the largest declines in poverty during the recent expansion, just as they were the main casualties of the prior economic contraction. Partly this difference results because black poverty was largely associated with joblessness, whereas Hispanic poverty derives mainly from low wages and low skills, which are less valuable in the current economy. Whether they or Hispanics will shoulder the consequences of the next recession remains to be seen, but the historical record is clear about the higher vulnerability of minorities to fluctuations in the business cycle, especially to labor market stresses triggered by industrial restructuring.

Getting Ahead: What Does It Take?

Before discussing the recent policy changes that altered welfare and work opportunities for poor urban populations, we return to where we began—

to the voices of Chicago's inner-city parents. Their aspirations and frustrations convey most forcefully what it takes to get ahead economically in Chicago, and their personal experiences tell much about the color of opportunity. Do inner-city minority parents perceive that group membership limits their life chances? And do they appreciate that the value of education has increased in recent years? In light of the rapid economic changes that drastically modified opportunities for earning a living, one might conjecture that inner-city parents are oblivious to what it takes to advance materially because their day-to-day struggles leave little room for planning, reflection, and action. However, their responses to a survey question about what it takes to get ahead in life (reported in table 7.2) and to open-ended questions asked in the Social Opportunity Survey reveal just the opposite.[3]

Their responses were highly consistent with our findings based on a statistical analysis of family, welfare, and work behavior. Virtually all parents underscored the necessity of *education* for getting ahead in life, and this response varied little by race, ethnicity, and gender. This sentiment, echoed throughout the Social Opportunity Survey, is illustrated by Rubén, an unemployed, married Puerto Rican father of five with twelve years of education and a highly unstable work history. Responding to the question of who gets ahead in Chicago, he commented,

> *Latin people have improved their situation, but they still aren't where they should be. . . . The person with good education [gets ahead]. The person who likes to work and has a good job and good opportunities. The Latinos always seem to be little bit behind. I don't know exactly why this is. I don't know if it is because we don't have enough education or what. We have a large percentage of our kids leaving school. They don't finish high school, and if you don't have a diploma at this time, it is really hard to get a diploma.*

Similarly, when asked who was most likely to get ahead in Chicago, Manuel, forty-three, a Mexican man (a married father of six who has nine years of schooling, lives in Pilsen, and has always held a job), replied,

> *The people that have the most education it's easy for them to find a job. If you have a skill, if you have been to a trade school—now more than ever you need something like that. The rest, forget it. If you don't have an education or you don't have a skill, forget it.*

3. The specific question was as follows: "Some people think that different things help them get ahead in life. How important do you think each of the following is in helping people get ahead? (Read list.) Would you say very important, somewhat important, or not important?" For parsimony, we report only the percent that said each factor was very important.

Table 7.2 Percent of Respondents Who Agree That Each Factor Is Very Important for Getting Ahead in Chicago: Men and Women Residing in Chicago's Inner City by Race and Ethnicity (Percent)

	Blacks		Whites		Mexicans		Puerto Ricans	
	Men	Women	Men	Women	Men	Women	Men	Women
Education	89.5	95.9	88.5	93.8	95.8	97.7	96.7	97.2
Money in family	51.8	56.5	52.3	48.4	38.6	35.0	47.2	35.6
Race	34.5	35.8	22.4	18.9	21.2	23.6	27.7	21.9
Hard work	71.3	69.7	85.2	84.7	73.4	77.5	67.8	67.8
Knowing right people	57.0	58.6	57.6	45.2	75.9	58.6	67.4	57.9
Luck	30.8	30.3	32.4	28.9	64.1	60.1	42.8	49.1
Neighborhood	23.2	23.7	20.1	18.6	46.1	46.3	33.6	36.8
Gender	15.0	16.6	13.6	17.6	26.0	22.4	27.2	17.3
Language	60.4	54.5	66.3	57.4	90.4	90.9	77.6	83.5
N	[308]	[719]	[127]	[237]	[228]	[261]	[148]	[306]

Source: UPFLS, 1987

Not surprisingly, blacks were more sensitive to the role of race in constraining opportunities for getting ahead. One out of three black respondents indicated that race matters for getting ahead compared to approximately one in five of nonblack Chicago parents. However, from this response, it is not obvious whether race is an asset or a liability, on average. Under some circumstances, being black can broaden opportunity, as for example, when affirmative action programs give special consideration to equally qualified minorities. But race most often limits alternatives in housing markets, in labor markets, and in a wide variety of social settings. Given the extensive evidence presented above about the disadvantaged labor market experiences of black relative to other inner-city parents, these responses regarding the significance of race most likely reflect the intense discrimination experienced by black parents. Responses to the SOS about access to opportunity affirm this interpretation.

Belinda, thirty-nine, a married black woman with six kids who had been on public aid since 1972, sees this quite clearly:

> *Well, like, I've been here all my life, you know, and it's a white-ruled system, you know. If you're white, you've got a better chance. And I've personally experienced that. . . . You know, I've went on jobs and seen a white person get hired and I was told there was no job. The job wasn't even, didn't even require skills, you know, things like that.*

But blacks are not the only ones aware of racial and ethnic prejudice in Chicago's labor market. John, thirty-five, a white father of one child holding a Master of Fine Arts degree from Chicago's prestigious Art Institute, put it bluntly when asked about who gets good jobs and who gets bad jobs. In his words:

> *What jobs? I guess if you're talking about being a short-order cook in a lot of diners, it helps to be Greek. If you're talking about pork barrel jobs in the city government, it helps to be black now [because Chicago had a black mayor then]. If you're talking about investment banking, it helps to be white.*

Many respondents—too numerous to report—spontaneously answered questions about opportunity and access to good and bad jobs with references to race and ethnicity. Some acknowledged that color could be used to reserve slots for minorities, but most agreed that the best jobs go to whites, especially educated whites, while the worst jobs go to blacks and Hispanics.

Brianna, forty-one, a legally separated mother of two, is considerably more emphatic about the role of racial discrimination in limiting opportunity for blacks. When asked who gets the best jobs in Chicago, she replied:

> *Well, I have . . . usually, so far as I see, it's the whites. I can't put it in so many words but . . . it's a lot of prejudice in Chicago, I know that much. And lot of times, the young whites that come out of school, most of their parents, I mean, they own their business and stuff like that. So they work for them, they've got their parents more so than a lot of black folks, Spanish, or whatever. So when they go out to look for a job, quite naturally they gonna look like, "Hey, this fellah's got his parents," and so and so, they did this and this. Quite naturally they'll hire him. And it's education-wise too. Now I noticed that a lot of blacks have been going back to school a lot, more than what they used to.*

Her insights about the signals employers use in hiring decisions are quite astute and consistent with social science arguments that, in addition to hard and soft skills, employers use social markers such as neighborhood and family status in determining who is likely to be a "good worker." Still, she acknowledges that education is crucial for a chance at the "good jobs."

The advantages conferred by families of orientation were also perceived to be important for subsequent generations to get ahead, as Brianna's discerning comment conveys. Chapters 4, 5, and 6 presented extensive evidence showing that disadvantaged family origins, such as having been reared on welfare or by a single parent, were key mechanisms perpetuating economic disadvantages over generations. Numerous Social Opportunity Survey respondents corroborated this idea. About half of black and white parents indicated that *family wealth* is very important for economic success. Lower shares of Mexicans and Puerto Ricans agreed that having money in the family was very important for getting ahead in life, but they were slightly more likely than blacks and whites to report that knowing the right people was essential for getting ahead in life, and especially that luck was important in getting ahead. Only about one-third of black and white respondents indicated that *luck* was very important for life chances compared to over 60 percent of Mexicans and close to 45 percent of Puerto Ricans.

About 70 to 80 percent of respondents said that *hard work* was essential for economic success, with Puerto Ricans agreeing least and whites agreeing most with this statement. However, except for Mexicans, relatively few inner-city parents agreed that *neighborhood* was consequential for long-term life chances. Less than one in four black respondents concurred that where one lived affected prospects for getting ahead in life, and even lower shares of white parents did. This is surprising because, as we showed, blacks live in much poorer neighborhoods than whites, and, for those with limited access to transportation, job opportunities are circumscribed by place. Nearly half of Mexicans agreed that where one lives affects economic prospects, probably because they witness the contrast in job availability as they travel to work outside their neighborhoods in far greater numbers than blacks.[4]

Both *language* and *gender* were salient factors, limiting life chances more for Mexicans and Puerto Ricans than for whites and blacks. About 60 percent of white and black parents indicated that language inadequacies limited chances for getting ahead compared to 90 percent of Mexicans and 80 percent of Puerto Ricans. The high immigrant composition of our Mexican sample helps understand their response to the significance of language for economic success, but this does little to explain why higher shares of Mexican and Puerto Rican men relative to white and black men and women believe that gender limits their opportunities. Most likely these differential responses about gender reflect cultural differences in sex roles between blacks and Mexicans. As a largely foreign-born population, Mexican women are more traditional with respect to activities outside the home.[5]

Most inner-city parents recognize the value of having good connections in order to advance economically. Roughly 60 to 75 percent of inner-city respondents said that knowing the right people is essential to get ahead in Chicago. Having networks and connections was another recurrent theme in the open-ended responses to questions about access to good and bad jobs, but no one clarified how one gets connections in the first place, except that *the rich and white folks have them.* In the words of Susan, a white, welfare-reliant mother of three with eight years of formal schooling, you need to know "people that know other people in high places. If you know someone in high places, then they could maybe get them a job. If you know someone up there, then you would be able to get a job." Toribio, a Mexican immigrant with two years of formal schooling, reinforces arguments about the importance of networks for getting a foothold in the U.S. labor market, which socially isolated black men lack. When asked about access to good and bad jobs, he commented, "If you don't

4. The growing Hispanic presence is felt throughout the city because they occupy low-skill jobs in growing numbers in neighborhoods where they do not reside as well as their residential neighborhoods.

5. This was evident in the administration of the SOS interviews, which were often conducted jointly with both spouses because many Mexican husbands insisted on being present during the interview.

have friends there to put in a word for you, forget it. You need friends that have good jobs who can tell you when there are jobs and who can put in a word for you. They can recommend you and tell them that you are a good worker."

Yet, compared to the value of education, our respondents considered having connections, living in the right neighborhood, and racism relatively less important for gaining access to economic opportunity in Chicago, as figure 7.1 shows. This profile of responses is highly consistent with the story about color and opportunity based on our analysis of the survey data. That is, we demonstrate that educational deficits carry lifelong consequences for economic well-being primarily because workers lacking high school diplomas or college simply cannot compete well in the labor market and often spend prolonged periods of time jobless and welfare-reliant. All of the other factors considered matter less than education for improving opportunity except, perhaps, the availability of jobs. This is the second most important theme about opportunity, namely, that its *contours* depend heavily on the condition of the local and regional economy, as demonstrated by comparisons between Chicago and all central cities. That is, if the *color of opportunity* depends on the early life experiences and if pivotal life-course events differentially prepare black, white, and Hispanic youth for adulthood, the *contour of opportunity* depends on the tightness of the labor market and income-maintenance safety nets. Both of these contours—the availability of jobs and income supports for the poor—

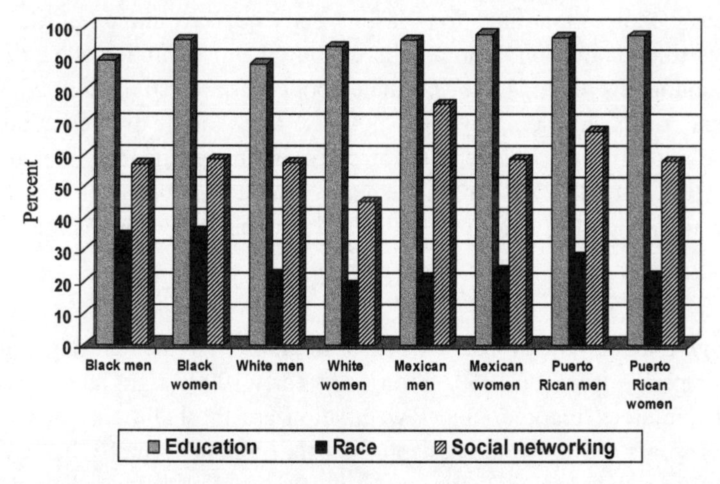

Figure 7.1 Percent of Respondents Who Agree That Each Factor Is Very Important to Getting Ahead in Chicago

have changed dramatically since 1987, when our respondents were interviewed. We close with a discussion of these trends and their implications for the contours of future inequality.

Epilogue: Reflections on Recent Economic Trends and Policy Reforms

Because the vigorous growth of the 1990s has been associated with the lowest unemployment rates in decades, it is useful to ponder the relevance of our findings for current and future policy discussions about the economic well-being of families in the lowest quintile of the income distribution. Several questions suggest themselves. First, what has happened to unemployment rates, and have racial and ethnic differences converged, or does group membership qualify labor market opportunities both in good and bad times? Second, how has Chicago fared in the most recent wave of economic growth? Has it been infused with new economic opportunities, or have the suburbs continued to grow at the expense of the inner city? And what does this portend for racial and ethnic differences in access to jobs in light of the continued changes in the ethno-racial ecological landscape? Third, how have welfare participation rates changed as a result of the recently enacted reforms? What does this portend for the future of the economically disadvantaged, in particular for the children of the poor who will face labor markets that require even more skills than they have in the past? In other words, will disadvantaged beginnings accentuate the contours of inequality along racial and ethnic lines in the future because of early choices that have long-lasting consequences into adult life, or is it possible to envision a future where group membership is not a correlate of economic standing? We speculate about these questions by reviewing recent trends in unemployment, poverty, and welfare participation and close with a brief consideration of corroborating evidence generated by the Multi-City Study of Urban Inequality conducted during the mid-1990s with the support of the Rockefeller and Russell Sage Foundation.

Does a Rising Tide Lift All Boats?

If high unemployment and slack labor markets were the defining features of the period from the 1980s through the early 1990s, tight labor markets, dynamic stock markets, and low inflation are the hallmarks of the last seven years of the twentieth century. This economic vitality has yielded promising dividends even for low-skill workers. Yet racial and ethnic differences in unemployment remain. In 1997, when national unemployment fell to 4.9 percent, down from 7.2 percent around 1985, black unemployment was twice the average in both years, 10 percent versus 15 percent, respectively (U.S. Department of Commerce 1998, table 646). Mexican and Puerto Rican unemployment also fell during this period, from 11.2 to 7.7 and from 14 to 9.8 percent, respectively—further reinforcing the old contours of stratification along racial and ethnic lines at the national level.

Despite the strong economic performance nationally, Chicago's unemployment rate was 7 percent in 1997—more than 2 percentage points above the national average—which implies even higher jobless rates for the city's minority workers, whose population share has continued to rise (U.S. Department of Commerce 1998, table 650). Because we show how early employment experiences influence subsequent adult outcomes in profound and decisive ways, we find it noteworthy that unemployment rates of minority youth also remained very high throughout the 1990s. In fact, unemployment among youth seems impervious to strong economic performance. In 1990, the youth unemployment rates were 19.9, 7.3, and 9.1 percent for blacks, whites and Hispanics, respectively. By 1997, the rates were 18.3, 6.9, and 10.3 percent for these same groups (U.S. Department of Commerce 1998, table 679). In short, a rising tide does not lift all boats, especially those of groups weighted heavily by the legacy of market discrimination (blacks) and educational deficits (Hispanics).

Furthermore, the benefits of economic growth were ecologically uneven: the robust economy produced far more jobs in the metropolitan area, while the city proper continued to lose jobs. That is, the trends presented in chapter 2 showing a polarizing of employment opportunities in the Chicago metropolitan area continued through the mid-1990s. From 1990 to 1996 the number of establishments covered by unemployment insurance increased unequally throughout the metro area, by 16.9, 11.9, and 8.2 percent, respectively, for the six-county metro area, Cook County, and Chicago. Furthermore, private-sector job growth for the total metropolitan area was 7.6 percent compared to only 2.8 percent in Cook County. However, during this period, the number of private sector jobs *fell* by 3.7 percent in Chicago proper (Illinois Department of Employment Security 1997, tables 1 and 2). Not only was job growth less impressive in Chicago during the robust economic recovery, but when the economy slows or goes into another recession, Chicago workers will most likely once again face more limited job opportunities than suburban workers. In short, recent trends indicate that the ecology of economic inequality continued to constrain income opportunities in Chicago even during a period of vigorous economic growth. Furthermore, pronounced residential segregation by race and Hispanic origin means that black, white, Mexican, and Puerto Rican residents confronted unequal job opportunities even during good economic times. That is, in addition to spatially delimited access to jobs for residents of large urban areas, persistent racial barriers also restrict employment options for inner-city workers.

But Chicago is not unique in affording unequal economic opportunities along racial and ethnic lines. Like the Urban Poverty and Family Life Survey, the Multi-City Study of Urban Inequality, which was conducted in Detroit, Boston, Los Angeles, and Atlanta during the mid 1990s, also found that space, skills, and race constrained economic prospects in these central cities.[6] The Multi-City Study concluded that while skills differentials and spatially distributed opportunities perpetuated inequality along racial and ethnic lines, racial barriers exerted the most pervasive influence on the employment options of inner-city workers. This conclusion, which is based on overwhelming evidence that blacks are less likely to be hired than whites, even in jobs with very low skill requirements (Holzer 1996), is entirely consistent with evidence reported for Chicago during bad economic times. Participants in the Multi-City Study conclude that America's deeply racialized social structures and institutions amplify the effects of past discrimination, making it difficult for race-neutral allocative mechanisms to transcend the contours of prior inequality along color lines. This generalization applies to our study about the color of opportunity in Chicago.

Welfare Reform and Income Maintenance

In 1996, Congress passed legislation that dismantled the Aid to Families with Dependent Children (AFDC) program and with it the primary safety net for poor families. The new welfare regime, Temporary Assistance to Needy Families (TANF), is putatively designed as a transitional solution to poverty, and thus prevents chronic dependence by precluding prolonged episodes of public assistance receipt. With limited exceptions, poor families cannot receive means-tested income support from the federal govern-

6. These observations are based on a prepublication newsletter issued by the Russell Sage Foundation to publicize the study. The Multi-City Study of Urban Inequality involved interviews with 8,600 African American, Hispanic, Asian, and white household members and 3,200 employers residing in the four cities. The study involved a uniquely linked set of surveys of employers and households and provides a detailed picture of the interacting dynamics of labor markets, residential segregation, and racial stratification in the four metropolitan areas.

ment for more than five years over an entire lifetime, and are required to work in exchange for their benefits after the first two years of public assistance. Though in the past the majority of poor families has used AFDC as a short-term source of support in difficult times as we documented in chapter 5, a nontrivial minority relied on public assistance for long-term support. This was especially so for residents of concentrated poverty neighborhoods, who are likely to have been reared in material deprivation.

Race is among the "risk factors" for long-term dependency, not because blacks and Hispanics are more predisposed to use welfare or because they prefer economic dependency to gainful employment, but because minority mothers are at greater risk of having been raised poor, of having truncated their educational careers, and of having entered family life via births rather than marriage. These early life-course decisions that are correlated with race and Hispanic origin rather than group membership per se are what undergird the higher risk of welfare receipt among minority women, especially black women. Other characteristics associated with long-term receipt include nonmarital childbearing, especially at young ages, low education, and limited or no work experience—the very attributes we identified and examined for Chicago's inner-city poor. That blacks and Hispanics are more likely than whites to have these characteristics implies that the impact of welfare reform will take its heaviest toll among them.

It is too early to report on the consequences of the new welfare legislation for minority families because the term limits are only now being reached in states that implemented it early. However, trends in poverty rates expose the higher vulnerability of people of color. Between 1970 and 1990, the national poverty rate fluctuated in accordance with the business cycle, but it rose by 1 percentage point (from 12.6 to 13.5 percent) over the twenty-year period. Throughout the period, racial and ethnic differences in poverty were quite pronounced. In 1970, the white poverty rate was just under 10 percent, compared to 33 and 23 percent for blacks and Hispanics, respectively. The black poverty rate dropped by 1 point by 1990, while Hispanic and white poverty rates rose slightly, but the egregious differentials persisted. During the 1990s there was a crossover of Hispanic and black poverty, so that by 1996, 29 percent of Hispanics were below the poverty line compared to 28 percent of blacks (U.S. Department of Commerce 1998, table 756). Although white poverty also rose slightly during this period, minorities were just over 2.5 times more likely to be poor than whites. This differential has persisted for many years and seems unresponsive to policy or economic responses.

The crossover of Hispanic and black poverty is important for thinking about future trends in racial and ethnic economic inequality for several reasons. First, the resistance of Hispanic poverty to economic recovery has more to do with the nature of the jobs that Hispanics acquire than with whether or not they work. That is, Hispanic poverty is, in the main, working poverty, while black poverty also involves a large component of nonworking poverty. When the labor market tightens, as occurred during the latter part of the 1990s, black poverty declined as increasing numbers of poor blacks entered the labor force. However, Hispanic poverty, which is dominated by the experiences of Mexicans, involves high rates of labor force participation at very low wages, mostly because of their low educational attainment. This means that changes in welfare policy and income supports will not influence their economic status to the same extent as for blacks unless the wages of unskilled jobs can be supplemented with income supports to guarantee nonpoverty wages. Increases in earned income tax credits have alleviated economic hardship among the working poor to some extent, yet the low wages associated with working poverty cannot be totally offset by this highly successful income transfer program. Moreover, for the large number of undocumented immigrants in the United States (and Chicago), this program is not available without valid social security numbers for all family members.

Second, Hispanics are the most rapidly growing population. Population growth does not directly contribute to poverty, but the preponderance of unskilled workers among recent Hispanic immigrants, especially Mexicans, weakens their labor market position as a group, particularly in the current environment that places a premium on skills. U.S. immigration policy does not yet give highest preference to labor market skills, but even if it did, undocumented migration and the large volume of immigrants who enter as immediate family members of U.S. residents would continue to infuse a steady stream of unskilled Hispanic immigrants. In other words, Hispanic immigrants participate in the labor force at very high rates, their very low skills place them at a severe disadvantage vis-à-vis blacks, which is evident in the trends in poverty rates, as well as changes in the ethnic composition of the poor. As the Hispanic population swells, the share of the poor who are Hispanic will also increase. Therefore, welfare policy that simply aims at making poor people work for their meager monthly check can not begin to address racial and ethnic differences in poverty unless the persistent problem of low wages is confronted head on.

This raises a third and crucial issue for the future, namely, that the trend in child poverty along racial and ethnic lines also bodes ill for the future contours of inequality and opportunity. Between 1990 and 1996, racial differences in poverty converged as the black rate fell from 44.2 to 39.5 percent, while white child poverty rose very slightly, from 15.1 to 15.5 percent. During the same period, Hispanic child poverty rose over 2 percentage points from 37.7 to 39.9 percent (U.S. Department of Commerce 1998, table 757). The proportion of poor children covered by AFDC also rose steadily from 1970 to 1990, from 59 to 67 percent (U.S. House 1998, table 7.10). However, as the strict time limits implemented as part of the new welfare legislation go into effect, the number of children who benefit from income transfers will decline dramatically. Evidence of the consequences of the new welfare legislation for women and children is beginning to emerge as we close our volume. If welfare agencies are successful in placing former welfare recipients into gainful employment during prosperous economic times, two policy questions remain. Will similar success rates continue when the economy slackens? Past experience suggests the answer is no. Second, and most important, do the jobs secured by former welfare clients lift families above the poverty threshold once the income transfer packages are assembled? The answer to this question appears to be no, even in good economic times. What opportunities can we expect for the minority children whose early beginnings imply serious material disadvantage? The lessons from the Urban Poverty and Family Life Survey are sobering at best.

More generally, if the contours of inequality remain intact during periods of economic vitality, it is worrisome to ponder what will happen when the economy cools off and unemployment rises. The evidence presented in this volume, when weighed against recent evidence from the Multi-City Study and recent trends in unemployment, poverty, and welfare participation, suggests that the color of opportunity could become even more divisive in the future.

APPENDIX A

Design of the Urban Poverty and Family Life Survey

The Urban Poverty and Family Life Survey sample was drawn in three stages. First, using the 1980 census, the research team identified tracts with poverty rates of 20 percent or more. As shown in table A1, in 1980, 39 percent (330/854) of Chicago's census tracts had poverty rates of 20 percent or more, and therefore qualified for inclusion in the sample frame. The spatial distribution of poverty changed during the 1980s, as described in chapter 2, so that by 1990, 43 percent of Chicago's census tracts would have qualified for inclusion into the sampling frame using the 20 percent poverty criterion.[1] The individual block groups that compose these tracts were sorted into four racial/ethnic strata based on their proportional representation of blacks, whites, Mexicans, and Puerto Ricans, and a random sample of block groups was drawn from within each stratum. Of the 330 census tracts eligible for the sampling frame as of 1980, only 125 tracts actually contained block groups that were selected for inclusion in the second stage sampling.

As shown in table A1, ghetto-poverty neighborhoods are disproportionately represented among tracts included in the survey compared to tracts not selected into the sampling frame. However, owing to the ecological polarization process experienced in Chicago during the 1980s, (see chapter 2), some of the tracts selected for inclusion based on 1980 poverty status actually improved their economic status by 1990. Thus, 18 percent

1. The UPFLS sample was drawn in 1986, and the actual survey was conducted in 1987, which is much closer to the 1990 census.

Table A1 Distribution of Chicago Census Tracts, 1980 and 1990, by Poverty Rates (Percents)

	All Tracts		Tracts Not in Survey		Tracts in Survey	
Poverty Rate	1980	1990	1980	1990	1980	1990
0–4	20.0	21.0	23.0	24.0	0.0	2.0
5–9	19.0	14.0	22.0	17.0	2.0	2.0
10–19	22.0	21.0	24.0	22.0	12.0	14.0
20–29	15.0	15.0	11.0	13.0	41.0	28.0
30–39	13.0	12.0	10.0	9.0	29.0	27.0
40+	11.0	16.0	10.0	15.0	16.0	26.0
N	825	825	700	700	125	125

of tracts included in the sampling frame had poverty rates below 20 percent in 1990, compared to 14 percent in 1980. However, the share of tracts classified as ghetto-poverty areas increased even more dramatically during the period—from 16 to 26 percent. Obviously, the probability of selecting poor households is much greater in ghetto-poverty neighborhoods compared to working-class or affluent neighborhoods, but the high income segregation in Chicago means that even high-poverty tracts include block groups of affluent households.[2]

In the second stage, a sample of dwellings was randomly drawn and screened for interview eligibility based on age, sex, ethnicity, and parent status. For this stage, approximately twenty-one thousand dwelling units located in the one hundred block groups were drawn. Screening interviews were conducted with a household member aged fifteen or over in each selected dwelling unit. The interview solicited the names, sex, and ages of all household members, and for persons age fourteen and over, race/ethnicity, marital status, and number of children ever born and currently living in the household. Poverty status of households was *not* a criterion governing eligibility for the sample, which was focused on family status (i.e. parents), minority-group status, and age (18–44).

Because this screening procedure did not yield the target sample threshold of 500 each of Mexicans, Puerto Ricans, and non-Hispanic whites, an additional round of listing and sampling was required. This final stage

2. Two examples are Kenwood, which is adjacent to Hyde Park, home of the University of Chicago, and Uptown, an area experiencing rapid gentrification but also containing very poor block groups. So extreme is the income segregation in parts of Uptown that then-Governor Thompson's home was randomly selected into the sample. The Thompsons were screened out as ineligible on age criteria.

repeated the prior sample selection steps, except that dwelling units were selected disproportionately from block groups in low-income census tracts with high concentrations of Mexican, Puerto Rican, or non-Hispanic white residents. This round yielded an additional 17,803 dwelling units from 40 white, 17 Mexican, and 26 Puerto Rican block groups, of which 4,149 were screened. All eligible, non-Hispanic white parents were selected because the number identified in the previous round was well below the desired sample size. No blacks screened during this final wave were included in the sample because the desired quota had been filled in the initial screening.

Table A2 presents social and demographic characteristics of census tracts that were represented and those not represented in the final sample. Because the survey was not fielded until 1987—between two censuses—we present census data for both 1980 and 1990 to show how neighborhoods evolved over the decade the study was in progress. The first section of table A2 situates the sampled tracts within the neighborhood typology presented in chapter 2. Given that the sample was drawn predominantly from neighborhoods that were considered "poor" circa 1980, the preponderance of "underclass" neighborhoods in the sample is unexpected. Extremely poor neighborhoods represented 36 percent of all sampled tracts in 1980 and 49 percent in 1990. Over the decade a nontrivial share of the tracts included in the sampling frame "upgraded" by experiencing some form of gentrification, while others "downgraded" and became underclass areas. In other words, despite its focus on low-income and poor neighborhoods, the survey actually includes residents from a wide variety of neighborhood environments, and the mix of neighborhoods included in the sample frame became even more diverse over time. For example, in 1980, only 13 percent of the tracts included in the sampling frame corresponded to "stable middle class" or "gentrifying yuppie" neighborhoods, but by 1990 over one in four were so classified.

These changes occurred against a backdrop of a larger trend toward the social and economic polarization of Chicago neighborhoods detailed in chapter 2. A comparison of the characteristics of census tracts that were and were not included in the sample frame reveals many similarities. First, although the census tracts from which the sample was drawn lost population and their ethnic composition changed, similar changes occurred among the tracts that were not sampled. That is, the share of the white population fell from 46 to 39 percent among the tracts excluded from the sampling frame, and from 20 to 15 percent for those included in the sampling frame. In addition, mirroring the changes in citywide population

Table A2 Social and Demographic Characteristics of Chicago Census Tracts, 1980–90 (Means or Percentages; standard deviation in parentheses)

	1980		1990	
	Tracts Not in Survey	Tracts in Survey	Tracts Not in Survey	Tracts in Survey
Neighborhood Type				
Stable middle-class	46.0	7.2	38.4	7.2
Gentrifying yuppie	11.9	5.6	21.4	20.0
Transitional working-class	21.8	51.2	12.3	24.0
Underclass	20.3	36.0	27.9	48.8
Tract Characteristics				
Population	3,403.0	4,415.5	3,202.3	3,857.3
	(2,396.9)	(2,745.9)	(2,377.4)	(2,357.6)
% White	46.5	20.3	39.3	15.3
	(38.6)	(21.3)	(36.0)	(20.1)
% Hispanic	12.7	30.9	16.6	32.5
	(19.0)	(31.5)	(23.1)	(34.6)
% Black	37.7	46.9	40.7	50.5
	(44.4)	(43.7)	(44.5)	(42.6)
% Foreign-born	13.5	16.9	13.2	15.6
	(12.5)	(16.7)	(13.1)	(15.8)
% College grads	12.7	7.2	22.5	14.8
	(14.6)	(9.1)	(19.2)	(13.4)
% Poor families	16.9	30.1	19.5	32.7
	(16.8)	(11.2)	(19.9)	(15.8)
% Public assistance	16.1	25.7	17.0	26.9
	(16.3)	(13.5)	(17.2)	(15.1)
Male LFPR	70.0	67.0	69.9	67.0
	(13.8)	(10.9)	(14.1)	(12.5)
Female LFPR	49.9	44.2	55.4	49.4
	(12.8)	(9.7)	(13.6)	(10.3)
Total LFPR	59.3	55.1	62.2	57.9
	(12.1)	(9.7)	(12.7)	(10.4)
Male unemployment rate	11.6	14.9	15.4	18.7
	(9.1)	(7.2)	(14.2)	(11.8)
Female unemployment rate	9.9	15.2	13.2	17.5
	(8.7)	(8.2)	(12.7)	(10.1)
Total unemployment rate	10.9	14.9	14.1	18.0
	(8.0)	(6.6)	(11.8)	(9.6)
% Owner occupancy	40.5	23.8	42.7	25.7
	(25.2)	(13.1)	(25.2)	(14.0)
% Same house five years	58.5	53.4	55.7	52.8
	(15.9)	(13.7)	(15.6)	(11.0)
N	700	125	700	125

Source: 1980 STF-3A and 1990 STF-3A

reported in chapter 2, the Hispanic share of all tracts increased. However, minorities are more highly represented among the sampled tracts—a deliberate feature of the stratified sampling design in a highly segregated city.

Second, socioeconomic differences evolved in parallel fashion in both groups of tracts, although sampled tracts were poorer, on average. Poverty rates rose throughout Chicago between 1980 and 1990 by roughly equivalent levels (i.e., roughly 3 percentage points, on average) in sampled and nonsampled tracts. The share of residents who received public assistance followed a similar pattern. By design, the average share of poor families is higher—approximately twice as high—in the tracts sampled than in those not represented in the sample, but this difference narrowed over the decade. Differentials in men's labor force participation rates were relatively small between tracts represented and not represented in the survey. Unemployment rates reveal much greater hardship in the tracts included in the sample than in those not represented. Finally, sharp differences exist in the owner occupancy rates between the two groups of census tracts.

In sum, census tracts included in the sampling frame are more disadvantaged than those excluded, on the basis of poverty and unemployment rates, but the citywide trends in poverty, unemployment, and population composition during the 1980s were similar in direction if not in magnitude. Minorities comprised larger population shares of the represented tracts because the sample design required stratification by race and Hispanic origin. Thus, the survey respondents are likely to face greater hardship than Chicago residents overall, and the UPFLS is representative of low-income neighborhoods.

Data Quality

Comparisons of data from the first-phase screening survey to the 1980 census revealed that UPFLS sampling procedures were adequate for the desired sample. The 1986 first-phase screening survey produced a percentage distribution of poverty tract population by race and Hispanic origin that was comparable to that based on the 1980 census. This survey showed that about 66 percent of poverty tract residents were black, and 21 percent were Hispanic. Also the screening survey and the 1980 census produced a consistent sex ratio, namely, 83:100 for the survey versus 84:100 for the census. Estimates of the proportion of men and women who had a child by age eighteen based on the screening survey were similar to those based on the 1979 National Longitudinal Survey of Youth (Krogh 1991). Finally, a comparison of the marital status distributions between UPFLS respondents and black men residing in the Chicago SMSA surveyed in the 1987–89 Current Population Survey revealed striking similarities (Testa and Krogh 1990). These comparisons provide great assurance that the UPFLS adequately represents the target population.

Sample Weights

One final point concerns the way we adjust for the stratified sampling design in the analyses we present throughout the book. Probabilities of selection differed among demographic groups; therefore, statistical analyses require weights to ensure statistical representation of the citywide population residing in census tracts with poverty rates of 20 percent (as of 1980). Weights provide unbiased estimates of means and proportions, but they do not provide accurate standard errors from multivariate analyses. Therefore, we weight all descriptive statistics, but we do not employ weights in any of the multivariate analyses. Sample *N*'s reported with weighted tabulations are not weighted, however.[3] Finally, because observations were randomly selected within sampling strata, our use of maximum likelihood methods produces unbiased and efficient estimates. This is critical for reliable statistical inference (Maddala 1983, 165–74).

3. During the early phases of the project, we conducted extensive sensitivity tests and determined that the design effects were quite small. Because our multivariate analyses were to model relationships rather than estimate population parameters, we did not use correction factors for design effects in any multivariate analyses. Again, the extensive sensitivity tests revealed that statistical inferences were unaffected by the decision not to weight multivariate analyses.

APPENDIX B

Methodological Appendix

This appendix discusses the technical aspects of data analysis and manipulation. First we describe the core predictor variables that were used throughout the analyses, including variation in operational descriptions for specific analyses. Subsequently, we discuss the techniques used in the analyses discussed in chapters 4, 5, and 6, respectively. We also report the methods used to convert logistic regression coefficients into probabilities that are depicted graphically throughout the text.

Table B1 presents the core independent variables and their operational definitions. Most analyses use three classes of predictor variables: individual attributes; family background characteristics; and structural characteristics. All models include race and ethnicity, either as predictors or as stratifying variables. Tabular analyses of the national sample of urban parents exclude Puerto Rican men as a separate group due to the small number of cases in the NSFH. Age was used either as a continuous variable or grouped into three discrete categories (18–24; 25–34; and 35–44 as the reference group). In addition, age served as the basis for creating the person-year files, which we elaborate below. Where appropriate, statistical models include controls for respondents' marital status at the time of the survey.

Family Background

Four variables describe respondents' family of orientation, and these are presented under the category of family background in table B1 and

Table B1 List of Independent Variables Included in the Analyses

Variable Name	Operational Definition/Variable Categories	Question Asked/Variable Construction
Individual Characteristics		
Race and ethnicity	Black White Mexican Puerto Rican	"Do you consider yourself of Mexican or Puerto Rican descent? If yes, which one?" "What race do you consider yourself to be, white or black?"
Age	Years: range from 18 to 44 (19 to 44 in the NSFH sample)	"When were you born?" and "How old are you now?"
Education	1 = High school or more 0 = Less than high school	"Did you graduate from high school?" (includes GED)
Marital status	Currently married Currently divorced Never married	"What is your current marital status? Is that: never married; married; married but separated; divorced; widowed?" Separated, divorced and widowed were grouped into "Currently divorced."
Family Background		
Family structure	1 = Both parents present 0 = one/no parent	"At what ages did you live with your mother?" (ranges birth to 21) "At what ages did you live with your father?" (ranges birth to 21)

Mother's education	1 = High school or more 0 = less than High school	"As far as you know, what is your mother's highest level of education?"
Family poverty status	1 = Parents ever received welfare 0 = Parents never received welfare	"As far as you know, during the time you were growing up until you were about 14 years old, did your family ever receive public aid?"
Welfare exposure	Long exposure Short exposure No exposure	"About how much of the time were they (your family) receiving public aid? Would you say almost all of the time, most of the time, about half of the time, some of the time, or almost none of the time?" Answers were grouped into three (1) all or most of the time (long exposure); (2) half or some of the time (short exposure); and (3) almost none or never.
Structural Characteristics		
Neighborhood poverty	Less than 20% 20–29% 30–39% 40% and above	Neighborhoods are defined as census tracts. Poverty was calculated as percent of families living in a census tract below the poverty line in 1980.
Neighborhood type	Stable middle-class Gentrifying yuppie Transitional working-class Underclass	The typology of neighborhoods is based on multidimensional array characteristics including: socioeconomic status, age composition, and residential stability (see chapter 2 for details).

throughout the text. Family structure indicates whether respondents were reared by one or both parents. We constructed this variable from information about the presence of mother and father during each year of life since birth until age fourteen. Family poverty status denotes whether respondents were reared in poor families, as indicated by having experienced episodes of welfare reliance when growing up. Some of the analyses used an alternative measure of poverty based on the length of exposure to welfare. We derived three exposure groups, which indicate whether respondents were exposed to public aid throughout their childhood (long exposure); whether respondents' families received welfare during part of their childhood; or whether they were never exposed to poverty. Parental socioeconomic status is proxied with mother's education. Father's education could not be used owing to excessive missing data. Because a relatively large number of cases lacked information for mother's education as well, we assigned mean values for tabular analyses. A flag for missing data on mother's education was added to all multivariate models to ensure unbiased parameter estimates.

Structural Characteristics

The main "structural" variables refer to neighborhood characteristics. Because the neighborhood characteristics were available only for a single point in time, we were unable to use them for the multivariate analyses based on the event history files. Instead, these attributes are used mainly as stratifying variables for various outcomes that were measured at the time of the survey or that did not change over time (i.e., race and ethnicity). Both of the neighborhood variables used, neighborhood poverty level and neighborhood type, are described extensively in chapters 2 and 3.

Methodological Approach

Multivariate analyses reported in chapters 4–6 employ a variety of statistical techniques, but specific applications differ in their details. In this section we describe the general procedures; subsequently we elaborate on key chapter-specific analyses.

The theoretical approach that guides our analyses, the life-course perspective, emphasizes turning points in life and focuses on both the occurrence of an event and its timing. The retrospective life histories provided in the UPFLS and the NSFH allow us to determine the timing of several life-course events on a yearly basis. They include marriage, births, high school graduation, employment, and welfare utilization. Therefore, our general methodological approach employs various forms of event history analysis (Allison 1988; Blossfeld, Hamerle, and Mayer 1989). We use models that employ both continuous and discrete time to analyze these life events. Continuous-time models are based on the duration of events, such as waiting time to marriage, job entry and exit, and length of unemployment and welfare spells. Discrete measures involve a person-to-person characterization of various statuses, including those that change yearly (e.g., children's ages). Examples of statuses that can change annually include marital status, employment status, or welfare use. The following sections discuss in detail the creation of the person-year file (discrete-time analyses) and the duration data files (continuous-time analyses).

The Person-Year Files

Most analyses of critical life events are based on discrete-time analyses. For these analyses we constructed a person-year file denoting whether a particular event occurred in a given year or at a specific age. Three person-year files were created and used in the analyses of pathways to family life, welfare participation, and labor force activity in chapters 4, 5, and 6, respectively. We describe each file in turn.

PATHWAYS TO FAMILY LIFE. To assess the pathways to family life, we computed a composite variable indicating whether a respondent experienced a marriage, a birth, or neither event in each year from age fourteen until age nineteen. Birth and marriage are treated as "competing" or exclusive events that lead to family life. This approach has the advantage of capturing both the timing and the sequencing of family events in a single measure. By the survey date, all respondents had experienced one or both events, because the survey was designed to include only parents. Because not all parents entered family life at the same age or in the same sequence, each individual contributes a different number of observations to the person-year file, depending on the age at which marriage or birth occurred. In other words, each individual contributed as many person-years as the years elapsed since turning fourteen until the time they formed a family, either via marriage or via a birth, or until they turned nineteen, whichever came first. Age nineteen was used as the end date for the construction of the pathways variable because of our interest in early disadvantage associated with teen parenting and premature assumption of adult roles (marriage). For example, a respondent who first married at age eighteen contributed five observations to the family-life file, while a respondent who did not marry but first had a birth at age sixteen contributed three observations. Respondents who did not experience their first birth or their first marriage until age nineteen contributed six observations to the person-year files. We used information on the dates of first marriage and first birth to calculate the timing of entry into family life.

WELFARE PARTICIPATION. To measure welfare participation, we constructed a person-year file denoting whether the respondent used (or did not use) welfare in each of the five years preceding the survey date. The UPFLS provides the exact dates for up to four welfare episodes, which were used to calculate recent and chronic use. However, the NSFH only records welfare use for up to five years prior to the survey date. Therefore, we constructed a person-year file that measures whether the respondent used welfare in each of the five years prior to the survey date. Each respondent thus contributed exactly five observations (person-years) to the file. This file is used to analyze "recent welfare use."

LABOR FORCE PARTICIPATION. To evaluate the employment returns to experience, we constructed a person-year file denoting whether respondents were in or out of the labor force at each age after they turned eighteen. Retrospective information that includes dates for all work episodes permits the construction of the multiple-observation data. Each respondent contributed person-year observations to the data from age eighteen until his or her age at the interview. The main advantage of a person-year approach is that we are able to consider the full work history of respondents rather than status at an arbitrary point in time (i.e., the time of the survey). This permits us to gauge the magnitude and consequences of employment instability over the life course.

Multivariate Models Based on Person-Year Files

The dependent variable in each specific person-year file is the occurrence of a particular event. The empirical models include two types of explanatory variables, namely, time-invariant and time-varying covariates. In all models we included several time-invariant covariates (see appendix table B1). Race and ethnicity is self-explanatory, as are respondents' education, age, and marital status. Family background includes several variables summarized in appendix table B1: (a) family structure; (b) mother's education; (c) family poverty status; and (d) welfare exposure. Depending on the particular application, we also introduced several respondents' characteristics, such as current age and current marital status.

The second class of independent variables included in the event history models are characteristics, such as education, marital status, employment status or age, that may change over time (time-varying covariates). Our statistical models include several time-varying covariates, some common to all models and others specific to particular analyses. The following time-varying covariates were common to all models: (1) respondent's educational level, which was depicted by high-school graduation status (1 = graduate) at each age; and (2) respondent's work status—a variable indicating whether the respondent worked prior to the beginning of each age interval. In all event-history models, we also include a control for age to monitor intra-interval variation in the timing of events. Several analyses included additional time-varying indicators, such as marital status or the age of the youngest child. We discuss these variables in the presentation of chapter-specific analyses.

Because all the dependent variables in the person-year files are measured in discrete time, we used logistic regression or, in some instances, multinomial logistic regression, as the basic statistical technique. The logistic regression estimates the log odds of experiencing an event, such as entry into family life or labor force participation (Maddala 1983). The probability of experiencing an event or, in the case of multinomial logistic regression, a set of events is calculated as follows:

$$P_j = \frac{\exp(\beta_j X_j)}{1 + \sum \exp(\beta_j X_j)}$$

where β_j denotes a set of parameter estimates and X represents a vector of theoretically relevant variables. For each model estimated, we transform the coefficients yielded by the (binomial or multinomial) logistic model into probabilities (Roncek 1991; Liao 1994). The procedure to translate logistic regression coefficients into probabilities is model-specific and is calculated for each independent variable of substantive and theoretical interest (e.g., ethnicity, level of education) while holding all other variables at their average level.

This approach weights the coefficients by the population-specific characteristics, because the means are derived from person-year files rather than individuals. Other alternative strategies to compute the predicted probabilities could be implemented, such as using a fixed profile of a person used consistently throughout. For example, a person raised by one parent, who was exposed to poverty as a child, had a high school diploma, and so on could be used to generate probabilities. Although this approach has the advantage of consistency, its main disadvantage is that the profile selected is totally arbitrary and probably fits no group well, if at all. Given that our focus is on group comparisons and that the groups differ appreciably in characteristics of theoretical importance, we chose to use group-specific average characteristics to represent average population profiles. This strategy builds on the fact that each of our models was based on a separate person-year file, with varying person years contributed by each respondent. For example, the number of observations in the "family pathways" file is determined by the age at which the respondent entered family life—either via marriage or via birth. Thus, some respondents will be censored after age sixteen, others after age eighteen, and so on. In contrast, the "labor force participation" file includes a complete set of observations corresponding to respondents' age at the time of the survey. That is, some respondents will contribute twenty person years, and others will contribute only two. We present the sample means that were used for calculations in appendix tables adjacent to the respective logistic regression results.

To illustrate this procedure with a concrete example, consider the procedure to calculate the probability that a black man in Chicago will drop out of high school (coefficients reported in table B4.1 below). First we calculated the log odds using the following equation:

$$\text{Logit}(P) = \alpha + \beta_i \overline{X}_i + \gamma_r$$

where α is the constant term, β_i refers to the ith coefficient of the independent variables in the model, excluding race and ethnicity, $\overline{X}$ denotes the mean value of these variables (see table B4.3), and γ is the vector for race and ethnicity (0.016 for blacks). By inserting the mean value of each of the additional covariates in the model, in effect we compare the probability for a black man with the average characteristics of the population in question to a white or Hispanic man with the identical characteristics. Using the same procedure, we derive the expected log odds for Mexicans, Puerto Ricans, and whites. For whites, $\gamma = 0$, because they are the reference group.

Second, after deriving the expected odds for all four race and ethnic groups, we convert them into probabilities using the following equation:

$$P = \frac{\exp(\theta)}{1 + \exp(\theta)}$$

where θ denotes the expected log odds for the specific group. These transformations were conducted on standard spreadsheets. In the example above, the log odds for Chicago black men are 1.424, and $p = .586$.

Duration Data File and Continuous-Time Analyses

Using a calendar, respondents provided detailed information on the dates when core events occurred in their lives. The surveys include complete information on marriage histories (i.e., dates of all marriages and divorces); fertility histories (i.e., the birth date of each child); employment histories, which include entry and exit dates for up to seven job episodes (and ten episodes in the national sample); and, for the UPFLS, entry and exit dates for up to four welfare spells. To measure duration, we used the beginning and ending dates for a particular episode of interest (i.e., job spells or welfare spells.) For right-censored cases (i.e., respondents who did not experience a particular event by the survey date), we assigned the date of the survey as the termination date.

Four duration measures were created: waiting time to family formation, analyzed in chapter 4; duration of welfare spells, analyzed in chapter 5; and duration of employment and unemployment spells, analyzed in chapter 6. Duration analyses were largely descriptive, in that we used life table methods to derive survival functions. The survivor functions reported in chapter 4 (presented in figures 4.4 to 4.8) are based on double-decrement life table estimates that portray the timing of marriage and births as competing risks to initiate family life. Life table analyses simulate the probability of an event (i.e., birth or marriage) occurring at each age interval, conditional on the competing event not having occurred by that age. The survivor functions indicate the hazard of not experiencing a particular event at a given age, from age thirteen to the survey date.

In chapter 5 we measure the duration of each welfare spell using the reported entry and exit dates (or the survey date for respondents who were still on a particular welfare episode at the time of the study). Thus, the reported survivor function shows the hazard of remaining on welfare for each year since beginning a spell.

Chapter 6 analyzes job exits and labor force reentry following a jobless episode. The duration of employment was measured for each job episode in the employment history using entry and exit dates. The survivor function reported in Figures 6.4 to 6.6 show the hazard of remaining in a job for each year since respondents first began a specific employment episode.

Job reentry measures the duration of joblessness starting at the date of job termination and culminating when a new job begins. The corresponding survival functions shown in figures 6.7 to 6.9 indicate the hazard of remaining jobless for each year since respondents exited the last work episode.

Chapter-Specific Methodological Appendices

In what follows we describe the specific multivariate analyses on which we base the graphs reported in chapters 4, 5, and 6. Each summary is accompanied by a table that reports the results of the multivariate analyses.

Chapter 4

ANALYSIS OF HIGH SCHOOL GRADUATION (NONGRADUATION). To evaluate the determinants of high school withdrawal (relative to graduation), we estimated a logistic regression in which the dependent variable indicated whether respondents *did not* graduate from high school by age nineteen. The dependent variable is computed using information about the respondents' birth date and the date of graduation. The independent variables, listed in table B4.1, included race and ethnicity and various indicators of family background, measured by family structure; number of siblings; mother's education; family poverty status; and age depicted as a five-category cohort measure (18–24, 25–29, 30–34, 35–39, and 40–44 as the reference group). The model also includes a control for missing data on mother's education. Table B4.2 presents the analyses of dropping out of high school for men and women in each of the three samples, and table B4.3 reports the sample means used to derive the predicted probabilities reported in figures 4.1 to 4.3.

PATHWAYS TO FAMILY LIFE. In any given year between ages fourteen and nineteen, the family status variable could assume three possible outcomes: having a birth, getting married, or neither. Therefore, we employed a multinomial logistic regression model to analyze pathways to family life. The independent variables include the time-invariant variables discussed above (namely, race and ethnicity and family background variables); two time-varying covariates (respondents' high school graduation status and work status in each year); and age. Results from empirical estimation are reported in tables B4.4 and B4.5 for women and men, respectively.

Table B4.1 Definitions of Specific Variables Analyzed in Chapter 4

Variable Name	Variable Categories	Operational Definition and/or Question Asked
Dependent Variables		
High school graduation	1 = Did not graduate 0 = Graduated	The variable indicates whether the respondent did not graduate from high school by age 19, based on the date of high school graduation.
Pathways to family	Entered family via marriage Entered family via birth Did not enter family = 0	Variable in person-year file indicated whether respondent had a birth for the first time or entered the first marriage (or none) in each year from age 14 until the experience of marriage *or* a birth, or until he/she turned 19, whichever came first.
Household income		
UPFLS:	10 categories, range between $1250 to $45,000	"Think about all the different types of income we've talked about. Last year, into what category did the total income fall for everyone living in your household?" Categories ranged between "less than $2500 to over $45,000. Mid-point value was assigned to each category.
NSFH:	Continuous measure	Actual annual income from all sources.
Current welfare use		
UPFLS:	1 = Currently on welfare 0 = Not on welfare	"What kind of aid is respondent currently receiving?" Coded 1 if receives any aid, 0 for none.
NSFH:	1 = Currently on welfare 0 = Not on welfare	If respondent received any income from public aid.
Specific Independent Variables for Pathways Analysis		
Education	1 = High school graduate 0 = Not high school graduate	Using the date of high school graduation (for graduates) we calculate for each year from age 13 whether the respondent already graduated.
Work	1 = Employed 0 = Not employed	On the basis of the job histories, we calculated for each year since age 13 whether respondents had a job.
Specific Independent Variables for the Analysis of Outcomes		
Family experience	1 = Married, entered via marriage 2 = Married, entered via birth 3 = Divorced, entered via marriage 4 = Divorced, entered via birth 5 = Never married	This variable combines the information on current marital status and the pathways to family life.

Table B4.2 Logit Estimates of High School Noncompletion by Age 19: Men and Women in Chicago and Urban United States (Asymptotic Standard Error)

	Men		Women		Urban U.S. Poor
	Chicago	Urban U.S.	Chicago	Urban U.S.	
Race/ethnicity					
Black	.016	−.003	.136	.019	−.408
	(.255)	(.206)	(.175)	(.148)	(.233)
Mexican	1.072*	.706*	1.106*	1.021*	1.559*
	(.372)	(.297)	(.277)	(.218)	(.387)
Puerto Rican	.943*	—[a]	1.012*	1.238*	1.167*
	(.349)		(.222)	(.378)	(.550)
Family Background					
Family structure	−.094	−.636*	−.266	−.763*	−.673*
	(.203)	(.181)	(.137)	(.131)	(.218)
Number of siblings	.060*	.114*	.072*	.083*	.055
	(.031)	(.033)	(.021)	(.022)	(.036)
Mother's education[b]	−.844*	−.118*	−.734*	−.180*	−.085
	(.218)	(.028)	(.143)	(.022)	(.036)
Family poverty status	.818*	.607*	.567*	.670*	.614*
	(.265)	(.243)	(.152)	(.161)	(.235)
Age Cohort					
18–24	−.760*	.658*	−.012	.271	−.762*
	(.354)	(.326)	(.226)	(.221)	(.390)
25–29	−.327	−.214	−.172	−.109	−.566
	(.308)	(.270)	(.208)	(.206)	(.376)
30–34	−.624*	−.346	−.238	.117	−.359
	(.297)	(.240)	(.201)	(.190)	(.369)
35–39	−.470	−.284	−.429*	.033	−.444
	(.293)	(.245)	(.210)	(.194)	(.398)
Model χ^2	106.8*	160.0*	178.9*	445.8*	143.0*
N	564	838	1234	1667	562

Source: UPFLS, NSFH

[a] Category excluded due to inadequate sample size.

[b] Model includes a control for missing on mother's education.

* $P < .05$

The probabilities reported in figures 4.7 to 4.9 were derived using the same procedure explained in the previous section. However, because the dependent variable in this analysis has three possible outcomes, the probability that each event will occur is calculated as follows:

$$P = \frac{\exp(\theta_1)}{1 + \exp(\theta_1) + \exp(\theta_2)}$$

Table B4.3 Means of Variables Used to Predict High School Dropout Probabilities: Women and Men in Chicago and Urban United States

	Men		Women		
	Chicago	Urban U.S.	Chicago	Urban U.S.	Urban U.S. Poor
Race/Ethnicity					
Black	0.49	0.22	0.54	0.27	0.35
Mexican	0.13	0.10	0.09	0.11	0.19
Puerto Rican	0.18	0.02	0.20	0.03	0.06
Family Background					
Family structure	0.59	0.68	0.53	0.63	0.51
Number of siblings	5.10	3.70	5.47	4.03	4.69
Mother's education[a]	0.35	0.59	0.31	0.51	0.37
Family poverty status	0.59	0.12	0.34	0.16	0.26
Age Cohort					
18–24	0.13	0.08	0.19	0.13	0.20
25–29	0.22	0.18	0.23	0.20	0.25
30–34	0.23	0.28	0.24	0.26	0.26
35–39	0.23	0.27	0.18	0.24	0.17

[a] Model includes a control for missing on mother's education.

Mean values of covariates used to derive the predicted probabilities are reported in appendix table B4.6.

CONSEQUENCES OF PATHWAYS TO FAMILY LIFE. To evaluate the consequences of alternative pathways to family life we examine several outcomes measured at the time of the survey. These include household income and welfare dependence (the definition of the two variables is reported in appendix table B4.1).

Household Income. The adjusted average household income presented in figure 4.10 was calculated using ANOVA, with household income as the dependent variable and family experience as the main independent variable. The means, adjusted for ethnicity, age and education, are presented in figure 4.12.

Welfare Participation. A similar model was used to derive the adjusted welfare participation rates. The dependent variable denotes whether respondents were on welfare at the survey date as a dichotomous outcome. Figure 4.13 presents adjusted values of welfare participation.

Table B4.4 Log Odds of Pathways to Family Life by Age 19: Chicago and Urban U.S. Women

	Chicago		Urban U.S.		Urban U.S. Poor	
	Marriage	Birth	Marriage	Birth	Marriage	Birth
Race/Ethnicity						
Black	−.331	1.269*	−.872*	1.124*	−.453	1.071*
	(.196)	(.182)	(.162)	(.150)	(.247)	(.255)
Mexican	.490*	.289	−.183	.469*	−.394	−.146
	(.223)	(.262)	(.176)	(.231)	(.238)	(.338)
Puerto Rican	.217	.528*	−.258	.776*	.134	−.764
	(.215)	(.214)	(.311)	(.324)	(.363)	(.757)
Family Background						
Mother's education	−.232	−.130	−.584*	−.103	−.338	−.141
	(.174)	(.113)	(.128)	(.149)	(.203)	(.266)
Family structure	−.074	−.412*	−.285*	−.684*	−.396*	−.544*
	(.147)	(.106)	(.121)	(.136)	(.178)	(.232)
Family poverty status	−.068	.511*	.066	.419*	.247	.620*
	(.170)	(.107)	(.165)	(.145)	(.213)	(.252)
Time-Varying Characteristics						
Education	−.231	−.526*	−.288	−.654*	.251	−.699
	(.275)	(.241)	(.183)	(.251)	(.326)	(.574)
Work	.329	.066	.279	.414*	−.218	.107
	(.210)	(.183)	(.174)	(.208)	(.275)	(.378)
Age	.582*	.536*	.657*	.530*	.610*	.547*
	(.049)	(.182)	(.043)	(.044)	(.062)	(.078)
Model χ^2	666.7		729.4		237.8	
α	<.001		<.001		<.001	
N[a]	6,572		7,565		2,083	

Source: UPFLS and NSFH
[a] Person-years
* $P < .05$

Chapter 5

Chapter 5 was devoted to the analysis of welfare participation, including current use (i.e., at the time of the survey); recent use (i.e., five years prior to the survey); and chronic use (i.e., welfare duration). The operational definition of the variables used in the chapter are presented in table B5.1. The measures of current welfare participation are used for descriptive purposes only; all multivariate analyses are based on the welfare event histories.

Table B4.5 Log Odds of Pathways to Family Life by Age 19: Chicago and Urban U.S. Men

	Chicago		U.S. Cities	
	Marriage	Birth	Marriage	Birth
Race/Ethnicity				
Black	.065	2.106*	−.502	.979*
	(.433)	(.472)	(.321)	(.384)
Mexican	.218	.336	−.101	−.744
	(.485)	(.651)	(.312)	(.778)
Puerto Rican	−.686	1.310*	—[a]	—[a]
	(.597)	(.532)		
Family Background				
Mother's education	−.719	.060	−.830*	−1.064*
	(.399)	(.227)	(.238)	(.438)
Family structure	−.072	−.015	−.486*	−.913*
	(.337)	(.210)	(.221)	(.375)
Family poverty status	−.146	.191	−.121	−.467
	(.427)	(.230)	(.324)	(.556)
Time-Varying Characteristics				
Education	−.390	−.557	−.428	.170
	(.636)	(.431)	(.313)	(.515)
Work	1.037*	−.027	.775*	.396
	(.337)	(.288)	(.230)	(.430)
Age	.656*	.768*	.911*	.661*
	(.120)	(.081)	(.105)	(.150)
Model χ^2	255.57		251.64	
α	<.001		<.001	
N[b]	3,494		4,276	

Source: UPFLS and NSFH
Note: Urban U.S. poor sample too small for reliable analyses of men.
[a] Category excluded due to inadequate sample size.
[b] Person-years
* $P < .05$

MULTIVARIATE MODELS OF WELFARE PARTICIPATION. We analyzed recent welfare use (see details above) using logistic regression analysis. The multivariate model of welfare use includes the following time-invariant covariates: race and ethnicity, family structure, mother's education (including an indicator for missing cases on education), exposure to welfare, and pathways to family life. The latter indicates whether the respondent entered family life via a birth or had neither married nor born a child in a specific year (entry via marriage is the reference category). Although all respondents had entered family life by the survey data, the timing and

Table B4.6 Means of Variables Used to Predict Pathways to Family Life Probabilities: Women and Men in Chicago and Urban United States

	Men		Women		
	Chicago	Urban U.S.	Chicago	Urban U.S.	Urban U.S. Poor
Race/Ethnicity					
Black	0.43	0.18	0.48	0.28	0.21
Mexican	0.22	0.16	0.14	0.11	0.21
Puerto Rican	0.17	0.02	0.20	0.03	0.04
Family Background					
Family structure	0.62	0.67	0.56	0.62	0.57
Mother's education[a]	0.30	0.59	0.29	0.53	0.41
Family poverty status	0.19	0.10	0.30	0.16	0.19
Time-Varying Characteristics					
Education	0.03	0.06	0.03	0.06	0.03
Work	0.09	0.11	0.06	0.06	0.07
Age	16.20	16.44	16.07	16.28	16.16

[a] Model includes a control for missing on mother's education.

"pathway" differ. Thus, at any specific age, respondents could assume only *one* of the following three statuses: neither parent nor married; parent; or married.

Five time-varying covariates were included in the model of recent welfare use, all of which pertain to the year preceding the index year. These time-varying covariates, described in table B5.1, are (1) whether the respondent had a preschool child at home in the previous year; (2) whether the respondent had already graduated from high school in the previous year; (3) whether the respondent was married in the previous year; (4) whether the respondent gave birth in the previous year; and (5) whether the respondent worked in the previous year. The statistical model also included the age at the prior year.

Tables B5.2 and B5.3, respectively, present results for these analyses for women and men. For women we conducted separate analyses for Chicago and both urban subsamples, but for men the urban poor sample was too small for reliable analyses of welfare users. The coefficients reported in tables B5.2 and B5.3 and the sample means reported in table B5.4 were used to derive the adjusted probabilities of welfare receipt presented in figures 5.1 to 5.7.

Table B5.1 Definition of Specific Variables Analyzed in Chapter 5

Variable Name	Variable Categories	Operational Definition and/or Question Asked
Dependent Variables		
Current welfare use	1 = Yes, 0 = No	Whether respondent was currently receiving welfare.
Recent welfare use	1 = Yes, 0 = No	Variable in person-year file, indicates whether respondent received welfare in each of the five years prior to the interview date.
Chronic welfare use	1 = Yes, 0 = No	Duration of welfare spell, calculated from entry and exit dates, and transformed into years, for UPFLS only. The questions were, "When did you start receiving your (first/next/current or most recent) public aid grant in your own name?" "When did this grant end?"
Specific Independent Variables		
Preschool child	1 = Yes, 0 = No	Using the fertility information (the year and month each child was born), the age of all children was calculated. For each of the five years prior to the study we calculated whether the respondent has a child five years of age or younger.
Education	1 = High school or more 0 = Less than HS	Using the date of high school graduation (for graduates), we calculated whether respondents had graduated by the previous year, for each of the five years prior to the interview.
Birth	1 = Yes, 0 = No	Using the fertility history, we calculated whether the respondent had a birth in the previous year for each of the 5 years prior to the interview.
Marriage	1 = Yes, 0 = No	Using the marriage history, we calculated whether the respondent was in a marriage in the previous year for each of the five years prior to the interview.
Work	1 = Employed 0 = Not employed	Using employment histories, we calculated whether the respondent had a job in the previous year for each of the five years prior to the interview.

Table B5.2 Logit Estimates of Recent Welfare Use: Women Residing in Chicago's Inner City and Urban United States

	Chicago	Urban U.S.	Urban U.S. Poor
Race/Ethnicity			
Black	.668*	.138	−.006
	(.087)	(.084)	(.103)
Mexican	−1.246*	.215*	−.064
	(.123)	(.109)	(.144)
Puerto Rican	.484*	.134	.007
	(.100)	(.394)	(.204)
Family Background			
Family structure	−.223*	−.173*	−.252*
	(.060)	(.077)	(.096)
Mother's education	−.161*	−.152	−.001
	(.068)	(.078)	(.014)
Welfare Exposure			
Long exposure	.304*	.390*	.367*
	(.082)	(.124)	(.151)
Short exposure	−.206*	.430*	.325*
	(.085)	(.094)	(.112)
Pathways to Family[a]			
Entered via birth	.373*	.253*	−.134
	(.066)	(.080)	(.109)
Neither birth nor marriage	−.139	−.462	−.261
	(.098)	(.129)	(.163)
Preschool child	1.010*	.622*	.384*
	(.066)	(.080)	(.096)
Respondent Characteristics			
Education	−.764*	−.645*	−.408*
	(.061)	(.076)	(.097)
Marriage	−1.991*	−1.192*	−.776*
	(.077)	(.077)	(.098)
Birth	.289*	.462*	.453*
	(.104)	(.109)	(.134)
Work	−1.569*	−1.063*	−.693*
	(.074)	(.073)	(.094)
Age	.013*	.005	.030
	(.005)	(.006)	(.007)
Model χ^2	3303.8	1372.7	281.1
α	<.001	<.001	<.001
N[a]	8,436	10,344	3,132

Source: UPFLS and NSFH

[a] Person-years

* $P < .05$

Table B5.3 Logit Estimates of Recent Welfare Use: Men Residing in Chicago's Inner City and Urban United States

	Chicago	Urban U.S.
Race/Ethnicity		
Black	1.326*	−.095
	(.212)	(.269)
Mexican	−.894*	−.776*
	(.292)	(.358)
Puerto Rican	.651*	.700
	(.241)	(.524)
Family Background		
Family structure	−.141	.032
	(.117)	(.247)
Mother's education	.042	−.494
	(.137)	(.233)
Welfare Exposure		
Long exposure	.823*	.866*
	(.153)	(.349)
Short exposure	−.074	1.304*
	(.214)	(.260)
Pathways to Family		
Entered via birth	.151	.072
	(.135)	(.301)
Neither birth nor marriage	−.120	.185
	(.192)	(.322)
Preschool child	.376*	−.101
	(.124)	(.273)
Respondent Characteristics		
Education	−.668*	−1.233*
	(.126)	(.232)
Marriage	−.415*	.117
	(.128)	(.242)
Birth	.126	.765*
	(.195)	(.290)
Work	−1.542*	1.263*
	(.142)	(.219)
Age	.051*	−.002
	(.011)	(.019)
Model χ^2	539.0	124.2
α	<.001	<.001
N[a]	4,177	5,288

Source: UPFLS and NSFH

[a] Person-years

$P < .05$

Table B5.4 Means of Variables Used to Predict Recent Welfare Use Probabilities: Women and Men in Chicago and Urban United States

	Men		Women		
	Chicago	Urban U.S.	Chicago	Urban U.S.	Urban U.S. Poor
Race/Ethnicity					
Black	0.40	0.20	0.47	0.25	0.33
Mexican	0.28	0.11	0.17	0.1	0.18
Puerto Rican	0.16	0.02	0.20	0.03	0.05
Family Background					
Family structure	0.63	0.82	0.56	0.78	0.73
Mother's education[a]	0.27	0.60	0.28	0.53	0.39
Welfare Exposure					
Long exposure	0.1	0.03	0.16	0.05	0.08
Short exposure	0.07	0.07	0.12	0.1	0.16
Pathways to Family					
Entered via birth	0.31	0.13	0.37	0.22	0.19
Neither birth nor marriage	0.23	0.27	0.18	0.2	0.15
Preschool child	0.59	0.48	0.55	0.46	0.56
Time-Varying Characteristics					
Education	0.41	0.86	0.45	0.79	0.63
Marriage	0.1	0.09	0.09	0.08	0.10
Birth	0.57	0.72	0.33	0.57	0.38
Work	0.47	0.86	0.28	0.56	0.38
Age	31.24	31.40	29.83	30.04	27.95

[a] Model includes a control for missing on mother's education.

Chapter 6

Chapter 6 analyzes various facets of labor force activity, including current employment status (at the time of the survey), respondents' job history, and wages associated with the current job. In addition to the dates of entry and exit, the survey instrument recorded information on various characteristics for each job, such as the size of the firm, the number of hours worked, whether the respondent was a union member, and—in the case of a job exit—the reason for leaving. We present the operational definition of these characteristics in appendix table B6.1. The NSFH work history was less comprehensive and included mainly the entry and exit dates for employment episodes. Because the NSFH employment information is much more limited compared to the UPFLS, the majority of employment analyses are restricted to the Chicago inner-city sample.

MULTIVARIATE MODELS OF LABOR FORCE PARTICIPATION. Labor force participation is analyzed using the person-year file that includes observations from age eighteen to the survey date for each individual. We use logistic regression to analyze the correlates of labor force activity in any given year based on this file. Independent variables predicting labor market outcomes include respondents' race and ethnicity and various measures of family background (family structure, mother's education, and family poverty status). Time-varying covariates that influence labor market outcomes also are modeled on a person-year basis. These include high school graduation and marital status in each year, and the presence of preschool children in any given year (see table B6.1). Of key substantive interest is our measure of prior labor force experience, which is measured as *cumulative* years spent at work prior to the index year. Using the logistic regression estimates presented in tables B6.2 and B6.3, respectively, for men and women in the three samples, we derived the probabilities presented in figures 6.1 to 6.6, using the conversion procedure elaborated above and the sample means reported in table B6.4.

ANALYSES OF WAGE RATES AND RESERVATION WAGES. A special feature of the UPFLS is that it includes direct measures of reservation wages for respondents who were jobless at the time of the interview. We use this information for wage analyses. In the multivariate analysis of wages, the dependent variable is the log of actual wages for employed parents. The wage regression includes two classes of predictor variables: (1) individual attributes, including race/ethnicity; job experience (actual number of years worked); respondent's education (measured as years of completed schooling); birthplace; and marital status; and (2) employment characteristics, including firm size, union membership, and full- vs. part-time work status. Table B6.1 reports detailed definitions of these variables. Since wages are observed only for those who work, the regression results based on this subsample could be biased due to systematic differences between those who work and those who don't. Therefore, we first computed the probability that respondents participated in the labor force. This analysis is based on a probit regression and presented in table B6.5. The model includes three types of variables (for definition see

Table B6.1 Definitions of Specific Variables Analyzed in Chapter 6

Variable Name	Variable Categories	Operational Definition and/or Question Asked
Dependent Variables		
Labor force participation	1 = Yes 0 = No	Variable in person-year file indicates whether respondent had a job in each year from age 18 until the interview date.
(Log) Current wage (for employed individuals)	(ln)$/hour	"What is your current wage or salary before taxes? Is that per hour, day, week, or what?" (Respondents were also asked how many weekly hours they work.)
Reservation wage (for jobless individuals)	(ln)$/hour	"What would the wage or salary have to be for you to be willing to take a job? Is that per hour, day, week, or what?" (Respondents were also asked how many days per week they want to work, and how many hours per day.)
Specific Independent Variables for Participation Analysis		
Labor force experience	In person-year file: years worked prior to specific age In wage analyses: number of years	Duration of employment spell, calculated from entry and exit dates, and transformed into years. The questions were, "When was the (first/next) time you had steady work, either full-time or part-time?" "When did you stop this steady job?"
Preschool child	1 = Yes 0 = No	Using the fertility information (the year and month each child was born) the age of all children was calculated. For each year since age 18, we calculated whether respondents had a child 5 years of age or younger.
Education	In person-year file: 1 = High school or more 0 = Less than high school	Using the date of high school graduation (for graduates) we calculated for each year since age 18 whether the respondent already graduated.
Reasons for job exit	Structural Individual	"What were the reasons this work ended?" The categories were: job ended; fired; quit; promotion; place of work closed; other reason. "Structural reasons" include: job ended, place of work closed.

Independent Variables for Wage Analyses		
Education	In years/grades	"What is the highest grade or year of regular school you completed and got credit for?" (ranges from 0 to 16)
Birthplace	1 = Foreign-born 0 = U.S.-born	"Was respondent born in the United States?"
Labor force experience	In years	Duration of employment spell, calculated from entry and exit dates, and transformed to commulative years since age 18. The questions were, "When was the (first/next) time you had steady work, either full-time or part-time?" "When did you stop this steady job?"
Married	1 = Yes, 0 = No	"What is your current marital status?"
Employment Characteristics		
Firm size	In four categories Less than 10; 10–99; 100–499; 500+ (reference)	"About how many people worked for your company, in all its branches and offices?"
Union membership	1 = Yes, 0 = No	"Were you a member of a union?"
Full- vs. part-time	1 = Part-time 0 = Full-time	"Did you usually work at least 35 hours a week?" (part-time = less than 35 hours)
Disability	1 = Yes, 0 = No	"Are you limited in the kind or amount of work you can do because of an impairment or health problem?"
Other adults	1 = Yes, 0 = No	Respondents were asked to report on every person in household. A household member, other than the respondent, aged 19 or over, was recorded as "other adult."

Table B6.2 Logit Estimates of Labor Force Participation: Men in Chicago and Urban United States (Asymptotic Standard Errors)

	Chicago	Urban U.S.	Urban U.S. Poor
Ethnicity			
Black	−.073	−.183*	.035
	(.067)	(.052)	(.142)
Mexican	−.052	.169*	.361*
	(.093)	(.080)	(.173)
Puerto Rican	.002	−.255	−.307
	(.088)	(.142)	(.254)
Family Background			
Mother's education	−.238	−.167*	−.385*
	(.062)	(.051)	(.153)
Family poverty status	−.033	−.198*	−.085
	(.034)	(.071)	(.142)
Family structure	.128*	−.199*	−.081
	(.055)	(.059)	(.137)
Time-Varying Human Capital			
Education	.466*	.960*	.706*
	(.055)	(.052)	(.125)
LF experience	.395*	.274*	.291*
	(.011)	(.008)	(.019)
Family Status			
Marriage	.572*	.845*	.371*
	(.059)	(.050)	(.132)
Preschool child	−.088	−.101	−.270
	(.063)	(.057)	(.150)
Age	−.140*	−.114*	−.147*
	(.007)	(.007)	(.017)
Constant	1.586	1.804	2.563
Model χ^2	2923.6*	3701.8	495.4
N[a]	8,777	14,777	1,780

Source: UPFLS, 1987

[a] Person-years

* $P \leq .05$

Table B6.3 Logit Estimates of Labor Force Participation: Women in Chicago and Urban United States (Asymptotic Standard Errors)

	Chicago	Urban U.S.	Urban U.S. Poor
Ethnicity			
Black	−.083	.241*	.100
	(.055)	(.037)	(.073)
Mexican	−.047	.070	.066
	(.093)	(.055)	(.095)
Puerto Rican	−.139	−.219*	−.363*
	(.072)	(.101)	(.170)
Family Background			
Mother's education	−.076	−.013	−.062
	(.049)	(.033)	(.071)
Family poverty status	.013	−.100*	−.318*
	(.019)	(.048)	(.081)
Family structure	.202*	.026	−.129
	(.044)	(.038)	(.071)
Time-Varying Human Capital			
Education	.696*	.911*	.709*
	(.045)	(.036)	(.070)
LF experience	.470*	.309*	.327*
	(.009)	(.005)	(.012)
Time-Varying Family Status			
Marriage	.156*	−.077*	−.015
	(.046)	(.032)	(.065)
Preschool child	−.233*	−.328*	−.281*
	(.048)	(.034)	(.069)
Age	−.119*	−.064*	−.067*
	(.005)	(.003)	(.007)
Constant	.243	−.140	−.041
Model χ^2	6266.0*	7530.4*	1768.5
N[a]	18,118	26,755	7,048

Source: UPFLS, 1987

[a] Person-years

* $p \leq .05$

Table B6.4 Means of Variables Used to Predict Labor Force Participation Probabilities: Women and Men in Chicago and Urban States

	Men			Women		
	Chicago	Urban U.S.	Urban U.S. Poor	Chicago	Urban U.S.	Urban U.S. Poor
Race/Ethnicity						
Black	0.47	0.19	0.26	0.54	0.24	0.32
Mexican	0.13	0.10	0.25	0.07	0.10	0.20
Puerto Rican	0.17	0.02	0.07	0.19	0.03	0.06
Family Background						
Family structure	0.62	0.83	0.78	0.56	0.80	0.75
Mother's education	0.33	0.59	0.38	0.29	0.51	0.35
Family poverty status	0.22	0.10	0.20	0.42	0.12	0.20
Human Capital						
Education	0.46	0.76	0.54	0.47	0.69	0.49
LF experience	4.65	5.96	4.63	2.51	3.54	2.39
Family Status						
Marriage	0.42	0.52	0.50	0.35	0.52	0.41
Preschool child	0.27	0.26	0.24	0.31	0.29	0.31
Age	26.80	27.00	26.08	26.40	26.70	25.40

table B6.1): (1) individual characteristics—race/ethnicity, high school graduation status, and an indicator for disability status; (2) family status—whether respondent has a preschool child, is married, and whether other adults are present in the household (as an index of constraints on time for child care); and (3) neighborhood characteristics—a measure of neighborhood poverty level. Use of neighborhood characteristics is appropriate for these models, which do not rely on retrospective measurement.

Using the results from table B6.5 we computed a selection correction term to represent systematic differences between parents who were in and out of the labor force at the time of the survey. This term, denoted lambda, is included in the wage equation as a predictor variable. The second-stage results are presented in table B6.6. From these results, we derive expected wages for *all* respondents, using the results from the regression equations based on the subset of employed parents. In table B6.7 we present the actual and predicted wages for men and women, by race and ethnicity. Respondents who had a job reported

Table B6.5 Probit Estimates of Labor Force Participation: Men and Women Residing in Chicago's Inner City (Asymptotic Standard Errors)

	Men	Women
Individual Characteristics		
Race/Ethnicity		
Black	−.403*	−.048
	(.203)	(.111)
Mexican	.573*	.403*
	(.265)	(.140)
Puerto Rican	−.059	−.227
	(.233)	(.125)
High school grad	.046	.119*
	(.024)	(.014)
Disabled	−.960*	−.759*
	(.174)	(.105)
Age		
18–24	−.233	−.581*
	(.245)	(.134)
25–34	−.127	−.286*
	(.141)	(.083)
Family Status		
Preschool child	.086	.009*
	(.166)	(.108)
Married	.400*	.311*
	(.137)	(.079)
Other adults present	.025	.315*
	(.137)	(.078)
Neighborhood Poverty Rate		
20–29	−.152	−.074
	(.221)	(.122)
30–39	−.396	−.068
	(.233)	(.134)
40+	.535*	−.618*
	(.258)	(.151)
Constant	.728	−1.178*
	(.398)	(.223)
Model χ^2	108.3	270.8
α	$<.001$	$<.001$
N	716	1,438

Source: UPFLS, 1987
* $p \leq .05$

Table B6.6 (Log) Wage Regressions for Men and Women Residing in Chicago's Inner City (Standard Error)

	Men	Women
Individual Characteristics		
Race/Ethnicity		
Black	−.384*	−.112
	(.093)	(.075)
Mexican	−.181	.082
	(.125)	(.121)
Puerto Rican	−.252*	.036
	(.110)	(.108)
Experience	.005*	.003*
	(.001)	(.001)
$(Exp)^2 * 10^{-2}$	−.009*	−.006
	(.003)	(.004)
Education (yr)	.044*	.066*
	(.010)	(.013)
Foreign-born	.090	.063
	(.098)	(.096)
Married	.088	−.008
	(.072)	(.061)
Job Characteristics		
Firm size <10	−.320*	−.436*
	(.088)	(.082)
Firm size 10–99	.019	−.075
	(.072)	(.069)
Firm size 100–499	−.113	−.079
	(.077)	(.076)
Union member	.164*	.173*
	(.063)	(.064)
Part-time job	−.055	−.208*
	(.088)	(.064)
λ	−.023	−.141
	(.217)	(.124)
Constant	1.16*	.873*
	(.223)	(.283)
R^2	.217	.253
N	593	663

Source: UPFLS, 1987
$* p \leq .05$

Table B6.7 Comparison of Actual, Expected, and Reservation Wages: Men and Women Ages 18–44 in Chicago's Inner City by Ethnicity in 1987 (Means or Percents)

	Black	White	Mexican	Puerto Rican
Men				
Employed				
(1) Actual wages	$8.32	$12.80	$7.35	$7.72
(2) Predicted wages	6.91	10.58	6.53	6.81
Jobless				
(3) Reservation wages	5.55	8.66	6.89	5.98
(4) Predicted wages	4.98	9.60	6.21	5.55
Women				
Employed				
(1) Actual wages	$6.77	$10.63	$4.74	$6.06
(2) Predicted wages	5.97	6.15	4.46	5.63
Jobless				
(3) Reservation wages	5.30	5.36	4.65	4.11
(4) Predicted wages	4.55	5.40	4.29	4.22

Source: UPFLS

their actual wages. Jobless respondents (both unemployed and nonparticipants) were asked about their "reservation wages"; that is, for how much money they would be willing to accept a job? We then took the ratio of the actual wage or reservation wage to the predicted wage derived from the estimates reported in table B6.7. The ratios are reported in figure 6.9.

References

Ahituv, Avner, and Marta Tienda. 1996. "Ethnic Differences in School Departure: Does Youth Employment Promote or Undermine Educational Achievement?" Pp. 93–110 in Garth Mangum and Stephen Mangum, eds., *Of Heart and Mind: Social Policy Essays in Honor of Sar Levitan.* Kalamazoo, MI: Upjohn Institute.

Ahituv, Avner, Marta Tienda, and V. Joseph Hotz. 1999. "The Transition from School to Work: Black, Hispanic, and White Men in the 1980s." Pp. 250–58 in Ray Marshall, ed., *Back to Shared Prosperity.* Armonk, NY: M. E. Sharpe.

Ahituv, Avner, Marta Tienda, and Angela Tsay. 1998. "Early Employment Activity and School Continuation Decisions of Young White, Black, and Hispanic Women." Unpublished manuscript, Princeton University.

Akerlof, George A., and Brian G. M. Main. 1981. "Unemployment Spells and Unemployment Experience." *American Economic Review* 70: 885–93.

Allison, Paul D. 1988. *Event History Analysis: Regression for Longitudinal Event Data.* Beverly Hills: Sage Publications.

An, C. B., R. Haveman, and B. Wolfe. 1993. "Teen Out-of-Wedlock Births and Welfare Receipt—The Role of Childhood Events and Economic Circumstances." *Review of Economics and Statistics* 75 (2): 195–208.

Anderson, Douglass K. 1993. "Adolescent Mothers Drop Out." *American Sociological Review* 58: 735–38.

Anderson, Elijah. 1991. "Neighborhood Effects and Teenage Pregnancy." Pp. 375–98 in Jencks and Peterson 1991.

———. 1990. *Streetwise.* Chicago: University of Chicago Press.

Año Nuevo de Kerr, Louise. 1975. "Chicano Settlements in Chicago: A Brief History." *Journal of Ethnic Studies* 2: 22–31.

Auletta, Ken. 1982. *The Urban Underclass.* New York: Random House.

Baca Zinn, Maxine. 1989. "Family, Race, and Poverty in the Eighties." *Signs* 14 (4): 856–74.

Bahr, Stephen J. 1979. "The Effects of Welfare on Marital Stability and Remarriage." *Journal of Marriage and the Family* 41 (3): 553–60.

Bane, Mary Jo, and David Ellwood. 1986. "Slipping into and out of Poverty: The Dynamics of Spells." *Journal of Human Resources* 21 (1): 1–23.

———. 1983. "The Dynamics of Dependence: The Roots to Self-Sufficiency." Report prepared for the assistant secretary for planning and evaluation, Office of Evaluation and Technical Analysis, Office of Income Security Policy, U.S. Department of Health and Human Services, Washington, DC.

Bean, Frank, and Marta Tienda. 1987. *The Hispanic Population of the United States.* New York: Russell Sage Foundation.

Becker, Gary. 1993. *Human Capital.* 3rd ed. Chicago: University of Chicago Press.

———. 1964. *Human Capital: A Theoretical and Empirical Analysis, with Special Reference to Education.* New York: Columbia University Press for NBER.

Ben-Porath, Yoram. 1967. "The Production of Human Capital and Life Cycle Earnings." *Journal of Political Economy* 75: 352–65.

Bianchi, Suzanne. 1995. "Changing Economic Roles of Women and Men." Pp. 107–54 in Farley 1995, vol. 1.

Blank, Rebecca. 1997. *It Takes a Nation: A New Agenda for Fighting Poverty.* New York: Russell Sage Foundation.

Blank, Rebecca M., and Patricia Ruggles. 1993. "When Do Women Use AFDC and Foodstamps? The Dynamics of Eligibility versus Participation." Unpublished manuscript, Northwestern University.

Blossfeld, Hans-Peter, Alfred Hamerle, and Karl Ulrich Mayer. 1989. *Event History Analysis: Statistical Theory and Application in the Social Sciences.* Hillsdale, NJ: Lawrence Erlbaum Associates.

Brooks-Gunn, James, Greg J. Duncan, and J. Lawrence Aber, eds. 1997. *Neighborhood Poverty: Context and Consequences for Children.* Vol. 1. New York: Russell Sage Foundation.

Brune, Tom, and Eduardo Camacho. 1983. *Race and Poverty in Chicago.* Chicago: The Chicago Reporter and the Center for Community Research Assistance, Community Renewal Society.

Cain, Glen G. 1985. "Welfare Economics of Policies toward Women." *Journal of Labor Economics* 3, part 2: 375–96.

Casuso, Jorge, and Eduardo Camacho. 1985. *Hispanics in Chicago.* Chicago: The Chicago Reporter and the Center for Community Research Assistance, Community Renewal Society.

Chicago Fact Book Consortium, ed. 1992. *Local Community Fact Book: Chicago Metropolitan Area.* Chicago: Chicago Review Press.

Clark, Kim B., and Lawrence Summers. 1979. "Labor Market Dynamics and Unemployment: A Reconsideration." *Brookings Papers on Economic Activity* 1: 13–72.

Clogg, Clifford C., Scott R. Eliason, and Robert Wahl. 1990. "Labor Market Experiences and Labor Force Outcomes." *American Journal of Sociology* 95 (6): 1536–76.

Coleman, James S. 1990. *Equality and Achievement in Education.* Boulder, CO: Westview Press.

Cutright, P., and H. L. Smith. 1988. "Intermediate Determinants of Racial Differences in 1980 United States Nonmarital Fertility Rates." *Family Planning Perspectives* 20 (3): 119–23.

Danziger, Sheldon H., and Peter Gottschalk. 1995. *America Unequal.* New York: Russell Sage Foundation.

———, eds. 1993. *Uneven Tides: Rising Inequality in America.* New York: Russell Sage Foundation.

Devine, Theresa, and Nicolas M. Keifer. 1993. "The Empirical Status of Job Search Theory." *Labour Economics* 1 (1): 3–24.

Donahoe, Debra, and Marta Tienda. 2000. "The Transition from School to Work: Is There a Crisis and What Can Be Done?" Pp. 231–63, in Sheldon H. Danziger and Jane Waldfogel, eds., *Securing the Future: Investing in Children from Birth to College.* New York: Russell Sage Foundation.

Drake, St. Clair, and Horace A. Cayton, eds. 1993. *Black Metropolis: A Study of Negro Life in a Northern City.* Chicago: University of Chicago Press.

Draper, Thomas W. 1981. "On the Relationship between Welfare and Marital Stability: A Research Note." *Journal of Marriage and the Family* 43 (2): 293–99.

Duncan, Greg J., Martha S. Hill, and Saul D. Hoffman. 1988. "Welfare Dependence within and across Generations." *Science* 239 (January 29): 467–71.

Duncan, Greg J., and Jeanne Brooks-Gunn, eds. 1997. *Consequences of Growing Up Poor.* New York: Russell Sage Foundation.

Duncan, Greg J., and Saul D. Hoffman. 1988. "The Use and Effects of Welfare: A Survey of Recent Evidence." *Social Science Review* 62 (2): 238–57.

Duncan, Otis Dudley, David L. Featherman, and Beverly Duncan. 1972. *Socioeconomic Background and Achievement.* New York and London: Academic Press.

Edin, Kathryn, and Laura Lein. 1997. *Making Ends Meet: How Single Mothers Survive Welfare and Low-Wage Work.* New York: Russell Sage Foundation.

Elder, Glen H. 1994. "Time, Human Agency, and Social Change: Perspectives on the Life Course." *Social Psychology Quarterly* 57 (1): 4–15.

Farley, Reynolds. 1995. *State of the Union: America in the 1990s.* 2 vols. New York: Russell Sage Foundation.

Featherman, David L., and Robert M. Hauser. 1978. *Opportunity and Change.* New York and London: Academic Press.

Fisher, Claude. 1995. "The Subculture Theory of Urbanism: A Twentieth Year Assessment." *American Journal of Sociology* 101: 543–77.

Fix, Michael, and Raymond J. Struyk. 1993. *Clear and Convincing Evidence: Measurement of Discrimination in America.* Washington, DC: Urban Institute Press.

Forste, Renata, and Marta Tienda. 1992. "The Schooling Consequences of Adolescent Sexual Activity." *Social Science Quarterly* 73 (1): 12–30.

Freeman, Richard B. 1992. "Crime and Unemployment of Disadvantaged Youths." Pp. 201–34 in George E. Peterson and Wayne Vroman, eds., *Labor Markets and Job Opportunity.* Washington, DC: Urban Institute Press.

Furstenberg, Frank F. 1976. *Unplanned Parenthood: The Social Consequences of Teenage Childbearing.* New York: Free Press.

Gottschalk, Peter. 1990. "AFDC Participation across Generations." *American Economic Review* 80 (May): 367–71.

Greenwood, Michael, and Marta Tienda. 1998. "U.S. Impacts of Mexican Immigration." Pp. 251–394 in *Binational Study: Migration between Mexico and the United States.* Austin, Texas: Morgan Press for the U.S. Commission on Immigration Reform.

Hannerz, Ulf. 1969. *Soulside: Inquiring into Ghetto Culture and Community.* New York: Columbia University Press.

Hirsch, Arnold A. 1983. *The Making of the Second Ghetto: Race and Housing in Chicago, 1940–1960.* Cambridge: Cambridge University Press.

Hodge, Robert W. 1973. "Toward a Theory of Racial Differences in Employment." *Social Forces* 52 (1): 16–31.

Hofferth, Sandra L., and Kristin Moore. 1979. "Early Childbearing and Later Economic Well-Being." *American Sociological Review* 44: 784–815.

Hogan, Dennis P., and Nan M. Astone. 1986. "The Transition to Adulthood." *Annual Review of Sociology* 12: 109–30.

Hogan, Dennis P., and Evelyn M. Kitagawa. 1985. "The Impact of Social Status, Family Structure, and Neighborhood on the Fertility of Black Adolescents." *American Journal of Sociology* 90 (January): 825–55.

Holzer, Harry J. 1996. *What Employers Want: Job Prospects for Less-Educated Workers.* New York: Russell Sage Foundation.

———. 1991. "The Spatial Mismatch Hypothesis: What Has the Evidence Shown?" *Urban Studies* 28 (1): 105–22.

Horan, Patrick M., and Patricia Lee Austin. 1974. "The Social Bases of Welfare Stigma." *Social Problems* 21 (5): 648–57.

Hotz, V. Joseph, Linxin Xiu, Marta Tienda, and Avner Ahituv. 1999. "Are There Returns to the Wages of Young Men from Working While in School?" Paper presented at the 1998 Meetings of the American Economic Association. Chicago, January 1998.

Hotz, V. Joseph, and Marta Tienda. In press. "Education and Employment in a Diverse Society: Generating Inequality through the School-to-Work." In Nancy Denton and Stuart Tolnay, eds., *American Diversity: A Demographic Challenge for the Twenty-First Century.* Albany, NY: SUNY Press.

Hsueh, Sheri, and Marta Tienda. 1996. "Gender, Ethnicity, and Labor Force Instability." *Social Science Research* 25 (1): 73–94.

———. 1995. "Earnings Consequences of Employment Instability among Minority Men." *Research in Social Stratification and Mobility* 14: 39–69.

———. 1994. "Race and Labor Force Instability." Pp. 95–100 in *Proceedings of the 1993 Annual Meetings of the American Statistical Association, Social Statistics Section.* Washington, DC: American Statistical Association.

Illinois Department of Employment Security. 1997. *Where Workers Work in the Chicago Metro Area: Summary Report: 1972–1996.* Chicago: State of Illinois.

Jargowsky, Paul A. 1996. *Poverty and Place.* New York: Russell Sage Foundation.

Jargowsky, Paul A., and Mary Jo Bane. 1991. "Ghetto Poverty in the United States, 1970–1980." Pp. 235–73 in Jencks and Peterson 1991.

Jencks, Christopher, and Paul E. Peterson. 1991. *The Urban Underclass.* Washington, DC: The Brookings Institution.

Kasarda, John P. 1995. "Industrial Restructuring and the Changing Location of Jobs." In Farley 1995, 1: chap. 5.

———. 1985. "Urban Change and Minority Opportunities." In Paul E. Peterson, *The New Urban Reality.* Washington, DC: Brookings Institution.

Kirschenman, Joleen, and Kathryn Neckerman. 1991. " 'We'd Love to Hire Them, but . . .': The Meaning of Race for Employers." Pp. 203–32 in Jencks and Peterson 1991.

Krogh, Marilyn. 1991. "A History and Description of the Chicago Urban Poverty and Family Structure Study." Unpublished background paper prepared for the Chicago Urban Poverty and Family Life Conference, September, 1991.

Lewis, Oscar. 1966. *La Vida: A Puerto Rican Family in the Culture of Poverty.* San Juan and New York: Random House.

———. 1961. *The Children of Sánchez: Autobiography of a Mexican Family.* New York: Random House.

———. 1959. *Five Families: Mexican Case Studies in the Culture of Poverty.* New York: Basic Books.

Liao, Tim Futing. 1994. *Interpreting Probability Models: Logit, Probit, and Other Generalized Linear Models.* Beverly Hills: Sage Publications.

Maddala, G. S. 1983. *Limited Dependent and Qualitative Variables in Econometrics.* Cambridge: Cambridge University Press.

Mare, Robert D. 1995. "Changes in Educational Attainment and School Enrollment." In Farley 1995, 1: chap. 4.

Massey, Douglas, and Nancy Denton. 1993. *American Apartheid.* Cambridge: Harvard University Press.

McLanahan, Sara S. 1988. "The Consequences of Single Parenthood for Subsequent Generations." *Focus* 2 (fall): 16–21.

———. 1985. "The Reproduction of Poverty." *American Journal of Sociology* 90: 873–901.

McLanahan, Sara S., and Karen Booth. 1989. "Mother-Only Families: Problems, Prospects, and Policies." *Journal of Marriage and the Family* 51 (3): 557–80.

McLanahan, Sara S., and Irwin Garfinkle. 1989. "Single Mothers, the Underclass, and Social Policy." *Annals of the American Academy of Political and Social Science* 501 (January): 92–104.

McLanahan, Sara S., and Gary Sandefur. 1994. *Growing Up with a Single Parent: What Hurts, What Helps.* Cambridge: Harvard University Press.

McLaughlin, S. D., B. D. Melber, J. O. G. Billy, D. M. Zimmerle, L. D. Winges, and T. R. Johnson. 1988. *The Changing Lives of American Women.* Chapel Hill: The University of North Carolina Press.

Menken, Jane. 1980. "The Health and Demographic Consequences of Adolescent Pregnancy and Childbearing." Pp. 157–200 in C. S. Chilman, ed., *Adolescent Pregnancy and Childbearing: Findings from Research, National Institutes of Public Health Publication No. 81-2077.* Washington, DC: U.S. Government Printing Office.

Miller, Brent C., and Kristin A. Moore. 1990. "Adolescent Sexual Behavior, Pregnancy, and Parenting: Research through the 1980s." *Journal of Marriage and the Family* 52 (November): 1025–44.

Mincer, Jacob. 1974. *Education, Experience, and Earnings.* Chicago: University of Chicago Press.

Moffit, Robert. 1992. "Incentive Effects of the U.S. Welfare System: A Review." *Journal of Economic Literature* 30 (March): 1–61.

Moore, Joan. 1989. "Is There a Hispanic Underclass?" *Social Science Quarterly* 70 (2): 265–84.

Murphy, Kevin M., and Finis Welch. 1993. "Industrial Change and the Rising Importance of Skill." Pp. 101–32 in Danziger and Gottschalk 1993.

Murray, Charles A. 1984. *Losing Ground: American Social Policy, 1950–80.* New York: Basic Books.

Nathanson, C. A., and Y. J. Kim. 1989. "Components of Change in Adolescent Fertility, 1971–1979." *Demography* 26 (1): 85–98.

O'Neill, June A., Laurie J. Bassi, and Douglas A. Wolf. 1987. "The Duration of Welfare Spells." *Review of Economics and Statistics* 69 (1987): 241–48.

Patterson, Orlando. 1993. "The Crisis of Gender Relations among African Americans." Unpublished paper, Harvard University.

PIChicago. 1995. "Chicago Employment Update." Unpublished paper, Chicago Private Industry Council.

Plotnick, Robert. 1983. "The Turnover in the AFDC Population: An Event History Analysis." *Journal of Human Resources* 18 (1): 65–81.

Popkin, Susan. 1990. "Welfare: Views from the Bottom." *Social Problems* 37 (February): 64–79.

Rank, Mark. 1994. *Living on the Edge: The Realities of Welfare in America.* New York: Columbia University Press.

———. 1988. "Racial Differences in the Length of Welfare Use." *Social Forces* 66 (4): 1080–1101.

Rein, Martin, and Lee Rainwater. 1978. "Patterns of Welfare Use." *Social Service Review* 52: 511–34.

Ricketts, Erol R., and Isabel V. Sawhill. 1988. "Defining and Measuring the Underclass." *Journal of Policy Analysis and Management* 7 (2): 316–25.

Rodman, Hyman. 1971. *Lower-Class Families: The Culture of Poverty in Negro Trinidad.* London: Oxford University Press.

Roncek, Dennis W. 1991. "Using Logit Coefficients to Obtain the Effects of Independent Variables on Changes in Probabilities." *Social Forces* 70 (2): 509–18.

Rosenfeld, Michael, and Marta Tienda, 2000. "Mexican Immigration, Occupational Niches, and Labor Market Competition: Evidence from Los Angeles, Chicago, and Atlanta, 1970–1990." Pp.64–105 in Frank D. Bean and Stephanie Bell Rose, eds., *Sociodemographic Implications of Immigration for African Americans.* New York: Russell Sage Foundation.

———. 1997. "Labor Market Implications of Mexican Migration: Economics of Scale, Innovation, and Entrepreneurship." Pp. 177–201 in Frank Bean, Rodolfo de la Garza, Bryan Roberts, and Sydney Weintraub, eds., *At the Crossroads: Mexican Migration and U.S. Policy.* New York: Rowman and Littlefield.

Ruggles, Patricia. 1990. *Drawing the Line: Alternative Poverty Measures and Their Implications for Public Policy.* Washington, DC: The Urban Institute Press.

Sampson, Robert J., and John H. Laub. 1993. *Crime in the Making: Pathways and Turning Points through Life.* Cambridge, MA: Harvard University Press.

SAS Institute, Inc. 1990. *SAS User's Guide: Statistics,* version 6, 1st ed. Cary, NC: SAS Institute.

Sewell, William H., and Robert M. Hauser. 1975. *Education, Occupation, and Earnings.* Ed. H. H. Winsborough. Studies in Population. New York and San Francisco: Academic Press.

Smith, James P., and Barry Edmonston, eds. 1997. *The New Americans: Economic, Demographic, and Fiscal Effects of Immigration.* Washington, DC: National Academy Press, 1997.

Spain, Daphne, and Suzanne Bianchi. 1996. *Balancing Act: Motherhood, Marriage, and Employment among American Women.* New York: Russell Sage Foundation.

Stack, Carol B. 1974. *All Our Kin: Strategies for Survival in Black Community.* New York: Harper & Row.

Stier, Haya, and Marta Tienda. 1997. "Spouses or Babies? Race, Poverty, and Pathways to Family Formation in Urban America." *Ethnic and Racial Studies* 20 (1): 91–122.

———. 1993. "Are Men Marginal to the Family? Insights from Chicago's Inner City." Pp. 23–45 in June C. Wood, ed., *Work, Family, and Masculinity.* Newbury Park, CA: Sage Publications.

Suttles, Gerald D. 1990. *The Man-Made City: The Land-Use Confidence Game in Chicago.* Chicago: University of Chicago Press.

Sweet, James A., and Larry L. Bumpass. 1987. *American Families and Households.* New York: Russell Sage Foundation.

Taylor, Robert J., Linda M. Chatters, M. Belinda Tucker, and Edith Lewis. 1990. "Developments in Research on Black Families: A Decade Review." *Journal of Marriage and the Family* 52: 993–1014.

Testa, Mark, Nan Marie Astone, Marilyn Krogh, and Kathryn M. Neckerman. 1989. "Employment and Marriage among Inner-City Fathers." *The Annals* 501 (January): 79–91.

Testa, Mark, and Marilyn Krogh. 1990. "Marriage, Premarital Parenthood, and Joblessness among Black Americans in Inner-City Chicago." Paper presented at the Joint Center for Political Studies and Department of Health and Human Services Forum on Models of Underclass Behavior, March 8–9, Washington, DC.

Tienda, Marta. 1995. "Latinos and the American Pie: Can Latinos Achieve Economic Prosperity?" *Hispanic Journal of Behavioral Sciences* 17 (4): 403–29.

———. 1991. "Poor People and Poor Places: Deciphering Neighborhood Effects on Poverty Outcomes." Pp. 244–62 in Joan Huber, ed., *Macro-Micro Linkages in Sociology.* Newbury Park, CA: Sage Publications.

———. 1990. "Welfare and Work in Chicago's Inner City." *American Economic Review* 80 (May): 372–76.

———. 1989. "Puerto Ricans and the Underclass Debate." *Annals of the American Academy of Political and Social Science* 501 (January): 105–19.

Tienda, Marta, and Leif Jensen. 1988. "Poverty and Minorities: A Quarter-Century Profile of Color and Socioeconomic Disadvantage." Pp. 23–61 in Gary D. Sandefur and Marta Tienda, eds., *Divided Opportunities: Minorities, Poverty, and Social Policy.* New York: Plenum Publishers.

Tienda, Marta, and Haya Stier. 1996a. "Generating Labor Market Inequality: Employment Opportunities and the Accumulation of Disadvantages." *Social Problems* 43 (May): 147–65.

———. 1996b. "The Wages of Race: Color and Employment Opportunity in Chicago's Inner City." Chap. 31 in Sylvia Pedraza and Rubén G. Rumbaút, eds., *Origins and Destinies: Immigration, Race, and Ethnicity in America.* Belmont, CA: Wadsworth Press.

———. 1991. "Joblessness or Shiftlessness: Labor Force Activity in Chicago's Inner City." Pp. 135–54 in Jencks and Peterson 1991.

Tienda, Marta, and Franklin Wilson. 1991. "Migration, Ethnicity, and Labor Force

Activity." Pp. 135–63 in John M. Abowd and Richard B. Freeman, eds., *Immigration, Trade, and the Labor Force.* Chicago: University of Chicago Press.

Upchurch, Dawn M., and Ames McCarthy. 1990. "The Timing of First Birth and High School Completion." *American Sociological Review* 55: 224–34.

U.S. Department of Commerce. 1998. *Statistical Abstract of the United States: 1998.* Washington, DC: Bureau of the Census, Economic and Statistics Administration, USGPO.

———. 1993. *1990 Census of Population. Social and Economic Characteristics.* Washington, DC: Bureau of the Census, Economic and Statistics Administration, USGPO.

———. 1982. *1980 Census of Population. Social and Economic Characteristics.* Washington, DC: Bureau of the Census, Economic and Statistics Administration, USGPO.

———. 1972. *1970 Census of Population. Social and Economic Characteristics.* Washington, DC: Bureau of the Census, Economic and Statistics Administration, USGPO.

U.S. House of Representatives, Committee on Ways and Means. 1987. *1987 Green Book.* Washington, DC: USGPO.

———. 1998. *1998 Green Book.* Washington, DC: USGPO.

Valentine, Charles A. 1968. *Culture and Poverty: Critique and Counter-Proposals.* Chicago: University of Chicago Press.

Vega, William. 1990. "Hispanic Families in the 1980s: A Decade of Research." *Journal of Marriage and the Family* 52: 1015–24.

Willis, Robert. 1986. "Wage Determinants: A Survey and Reinterpretation of Human Capital Earnings Functions" Pp. 525–602 in O. Ashenfelter and R. Layard, eds., *Handbook of Labor Economics.* Vol. 1. North Holland: Elsevier Science.

Wilson, Franklin D., and Lawrence L. Wu. 1993. "A Comparative Analysis of the Labor Force Activities of Ethnic Populations." *Proceedings of the 1993 Annual Research Conference, U.S. Bureau of the Census.* Washington, DC: USGPO.

Wilson, William J. 1996. *When Work Disappears.* New York: Knopf.

———. 1991. "Studying Inner-City Social Dislocations: The Challenge of Public Agenda Research." *American Sociological Review* 56 (February): 1–14.

———. 1987. *The Truly Disadvantaged: The Inner City, the Underclass, and Public Policy.* Chicago: University of Chicago Press.

———. 1979. *The Declining Significance of Race.* Chicago: University of Chicago Press.

Wilson, William J., and Kathryn M. Neckerman. 1986. "Poverty and Family Structure: The Widening Gap between Evidence and Public Issues." Pp. 232–59 in Sheldon H. Danziger and Daniel H. Weinberg, eds., *Fighting Poverty: What Works and What Doesn't.* Cambridge, MA: Harvard University Press.

Wirth, Lewis. 1938. "Urbanism as a Way of Life." *American Journal of Sociology* 44: 3–24.

Wu, Lawrence. 1996. "Effects of Family Instability, Income, and Income Instability on the Risk of a Premarital Birth." *American Sociological Review* 61 (3): 386–406.

Index

The letter *t* attached to a locator refers to a table; the letter *f* refers to a figure.

Aid to Families with Dependent Children (AFDC), 13–16, 132–33, 234; AFDC-U program, 123, 136, 148, 153; duration of spells, 16, 133, 152–55, 152f, 154f, 155f. *See also* Temporary Assistance for Needy Families (TANF); welfare participation

Astone, Nan M., 9

Auletta, Ken, 13n.6

Bane, Mary Jo, 8

birth pathway to family life, 111–13, 112f, 113f (*see also* teenage childbirth); age-specific factor in, 103t, 114, 128; effect on labor market options, 19, 21, 102; effects of childhood poverty on, 93–94, 104, 111–12, 118–19, 119f; among ethno-racial groups, 13, 112–14, 116f, 222–23; and future marital prospects, 109, 114; and high school noncompletion, 101, 109; household incomes associated with, 121–23, 121f; *vs.* national patterns, 13, 94, 102, 111–15, 112f, 113f, 119; as pathway to welfare use, 102, 109, 124, 132, 155–56, 155f, 235; place-specific differences in, 111–12, 112f, 113f; poverty effect on, 220; probabilities of, 114, 115–17, 116f, 124; race-specific factor in, 116–17, 128; significance of, as turning point, 3, 4, 10, 24–25, 101–2, 120, 127, 129, 145, 220, 222

"black belt," 35, 60–61. *See also* ghetto underclass neighborhoods

Black Metropolis (Drake and Cayton), 31, 35

black poverty, 6–7, 225, 236

blacks: birth pathway among, 94, 102–4, 103t, 112–14, 112f, 113f, 115–16, 116f, 222–23; and childhood exposure to poverty, 3, 95–96, 96t, 98, 220, 222; competition for low-wage jobs, 32, 40, 168, 212, 219; discrimination toward, 31, 63–64, 167–68, 182, 211–12, 215n.23, 222, 225, 234; educational levels of, 41t, 42, 84t, 102, 103t; family structure of, 3, 93, 94, 95; labor market participation of, 19, 43–44, 43t, 69, 84t, 85, 170–71, 171t, 193–95, 224; marriage pathway among, 3, 84t, 94, 110–11, 114f, 115, 146; migration to Chicago, 30–31; poverty among, 47, 69, 95–96, 141, 221; residence in ghetto neighborhoods, 36f, 37, 37f, 59, 66t, 68, 84t, 85, 141 (*see also* ghetto underclass neighborhoods); residential segregation of, 25, 31, 35, 36f, 37f, 138, 141, 149, 150, 163, 167, 177, 219, 222, 223 (*see also* jobs, spatial distribution of); and welfare participation, 46t, 47, 69, 84t, 85, 96, 126–27,

blacks (*continued*)
126f, 131–32, 134t, 135–41, 149, 153–55, 163, 220t, 222–23

Cayton, Horace A., 28–29, 31, 35
Chicago: contraction of population base in, 33–35, 40, 69; demographic trends in, 29, 30, 33–34, 34t, 69, 177; history of inequality in, 28–29; industrial restructuring of, 29, 47–52, 50f, 51f, 195; job decline in *vs.* in suburban jobs, 40, 48, 50f, 51, 51f, 177–78, 233; as national case study, 25–26, 78–79; spatial distribution of job decline in, 52–52, 177–78, 195, 233
Chicago Poems (Sandburg), 28
cluster analysis, 55, 56
competition, over jobs, 23n.14, 28–29, 168, 219
concentrated poverty, 7, 137, 149, 177
"concentration effects," 6, 106
"culture of poverty" thesis, 11, 12, 12n.5

Declining Significance of Race, The (Wilson), 5–6
Denton, Nancy, 8, 22n.13, 31, 31n.4, 177, 177n.6, 180n.8
disadvantage, transmission of: role of poverty in, 120, 220; role of race in, 120
disadvantages, labor market: educational deficits, 19–20, 40–42, 41t, 52, 69, 103t, 184; as result of life course events, 3, 8–9, 194–95; work experience deficits, 19–20, 69, 184–87, 185t, 186t, 213–14
discrimination. *See* job discrimination
divorce: economic penalties of, 121–22, 145; effect on economic well-being, 120–27
Drake, St. Clair, 28–29, 31, 35
dropout rates, from high school. *See* high school noncompletion

ecology, of neighborhoods: changes in, 30, 53, 60f, 61f, 62f, 68–70; types of neighborhoods within (*see* neighborhood typology). *See also* segregation, residential; neighborhoods, spatial polarization of
economic prosperity, and inequality, 2, 233
economy, 1980s *vs.* 1990s, 232–33
Edin, Kathryn, 161, 163, 199
educational attainment: effect of childhood poverty on, 93–94, 107–8, 107f, 184; effect of pathways to family life on, 23, 94, 101; effect on labor market options, 221–22, 231; ethno-racial differences in, 40–42, 41t, 84t, 85, 94, 102; and labor force participation, 190–91, 216
Ellwood, David, 8
employer preferences in hiring, 212, 215–17, 215n.23
employment instability: effect of early employment deficits on, 20, 168, 184–89; ethno-racial differentials in, 188, 196, 216–17; inner-city perceptions of, 200; *vs.* national patterns, 188–89, 196, 197f, 199f. *See also* labor market participation; work exit rates
employment opportunities: and discrimination, 163–64, 167, 211–12; and job availability, 168, 190–91, 190f; and labor market participation, 168; *vs.* national patterns, 4, 190–91, 190f
event-history methods, 139

family experience, 120–27
family life, pathways to: age-specific patterns, 117–18; among ethno-racial groups, 3–4, 12, 93, 94, 95, 102–4, 103t, 114; *vs.* national patterns, 13, 24, 94, 114–15, 222; place-specific effects on, 114–15, 222; as poverty effect, 117, 222–23; race-specific component to, 116f, 117, 119–20; related probability of future welfare use, 144–45, 145f, 235; significance of timing of, 19, 23, 109, 118, 221–22; and transmission of poverty, 93, 221. *See also* birth pathway to family life; marriage pathway to family life
family of orientation: economic well-being of, 94, 108, 148–49; group- *vs.* place-specific differences among, 95–96, 98; *vs.* national patterns, 93, 95–96, 96t, 98, 118–19, 119f, 136–37, 142–43, 143f; parental educational levels, 95; significance of family structure of, 94, 95–98, 96t, 106, 106f, 108, 117, 118t
family structure: as consequence of poverty, 11, 221; effect on educational attainment, 104, 106–7, 106f, 184; among ethno-racial groups, 3, 93, 95–98, 96t; influence of siblings, 96t, 97–98; and intergenerational family formation behavior, 117–18, 117f; and intergenerational transfer of disadvantage, 10–11, 93, 98, 108, 118, 120, 141–42, 221; *vs.* national patterns, 93, 95, 96t,

98, 117–18, 118f; and welfare use, 96–97, 141–42, 142f

General Assistance (GA), 132, 134, 152, 152f
ghetto underclass neighborhoods: economic deprivation within, 40, 53, 219; expansion of, 35, 36f, 37f, 61–66, 61f; joblessness within, 40, 167, 175t, 176–77; racial homogeneity of, 36f, 37, 37f, 59, 66t, 68, 84t, 85; and spatial distribution of jobs, 137, 150; welfare participation in, 59, 163
Growing Up with a Single Parent (McLanahan and Sandefur), 101

Hannerz, Ulf, 156–57, 156n.6
high school noncompletion, 104–8, 104f; effect of family structure on, 101, 106–7, 106f; among ethno-racial groups, 19, 40–44, 41t, 101, 102, 103t, 105; gender effect on, 105–6; *vs.* national levels, 102, 103t, 105; place-specific differences in probability of, 105–6, 108, 222; as poverty effect, 93–94, 104, 104f, 107–8, 107f; probability of, 104–8, 104f, 107f, 235; —, *vs.* national levels, 104f, 105; and probability of future welfare use, 143–44, 144f, 235; significance of, as turning point, 108, 221–22, 231
Hispanic poverty, 6–7, 90–91, 117, 225, 236–37
Hispanics. *See* Mexicans; Puerto Ricans
Hogan, Dennis P., 9
Holzer, 48, 215n.23
human capital investments: differential returns to, 21–22, 168, 183, 212, 213; and differential work experience, 4-5, 20, 21, 181, 191, 192f, 194, 213, 224; educational, 19, 181, 194, 213; among ethno-racial groups, 3, 4, 19–20, 21, 168, 213, 224; and return in weak labor market, 168; significance of timing of, 19; training, 20–21

immigrants, in labor force, 23n.14, 28–29, 32, 40
Immigration and Nationality Act (1965), amendments to, 32
income, adjusted household: absence of race effect, 125–26, 125f; gender-specific effect, 122–23; influence of family formation patterns on, 121f, 122; and marital status, 121–22, 121f; *vs.* national patterns, 122
incomes, median, 2n.1
index of dissimilarity, 40–44, 42n.6
inner-city residents (UPFLS respondents): economic disadvantages among, 85, 87, 137–38; educational levels of, 84t, 85, 107–8, 107f, 220t; incomes of, 121–22, 121f; labor force participation of, 43–44, 43t, 84t, 85, 87, 190–91, 192f, 193–95, 193f; neighborhood distribution of, 84t, 85; pathways to family life of, 83, 84t, 87, 220t (*see also* birth pathway to family life; family structure; marriage pathway to family life); percentage minorities, 86 (*see also* Chicago, demographic trends in); welfare use among, 84t, 85, 87, 220t (*see also* welfare participation)
intergenerational transmission of poverty, 14, 16n.9, 24–25, 98, 219, 220–22; family structure and, 98, 123; mechanisms facilitating, 98, 184

job discrimination: differential returns to human capital and, 21–22, 183; and ethno-racial wage inequality, 22, 163–64, 169, 208–9, 208t, 210f, 211, 225; and labor market outcomes, 19, 21; structural explanations of, 22 (*see also* queuing theory); toward blacks, 22–23, 31, 163, 167–68, 182–3, 211–12, 222, 225, 234
joblessness, 167, 170, 172, 174–76, 175t, 201–3; and availability of jobs, 190–91, 220; correlation with neighborhood poverty levels, 167, 213; and discrimination, 167, 216 (*see also* job discrimination); effect of early work deficits on, 20, 172–73, 233; among ethno-racial groups, 31, 167, 170, 171t, 174–76, 175t, 196, 216, 219; labor force exits and reentries and, 195–205, 197f, 199f, 201f, 202f; *vs.* national patterns, 167, 214; and permanent labor force withdrawal, 20; spatial distribution of, 174–76, 175t; among women, 167, 200
jobs: industrial *vs.* service, 48–51, 50f, 51f; job decline (*see* Chicago, job decline in *vs.* in suburbs); job turnover, 203; spatial distribution of, 29, 40, 47–53, 51f, 177–78, 195, 233 (*see also* Chicago, spatial distribution of job decline in)

labor market attachment, 7, 18–20. *See also* labor market participation
labor market behavior. *See* joblessness; labor market participation
labor market detachment, 172, 233. *See also* disadvantages, labor market
labor market instability: among minorities, 21, 201f, 202f, 204–5, 205f; *vs.* national patterns, 185t, 186t, 187; among women, 21, 187
labor market outcomes, differences in: and discrimination, 19, 163–64, 167, 211–12; and early work experience, 19, 20, 21, 168, 186–87; educational attainment and, 21, 168; factors leading to, 173, 182–83; neighborhood effect on, 173–77; queuing theory on, 22; and spatial mismatches, 173, 195; work-experience deficits and, 168
labor market participation, 170–74, 171t, 175t, 190–91, 192f, 193–95, 193f; correlation with welfare use, 47, 147–48, 147f; and education levels among groups, 41t, 42, 84t, 85; effect of discrimination on, 183, 194–95; among ethno-racial groups, 4, 19–20, 21, 43–46, 168, 170–74, 175t, 213, 224 (*see also* blacks; Mexicans; Puerto Ricans; whites); exits and reentries, 195–205; importance of early experience for, 4, 19–20, 221–22, 224, 233; job availability as factor, 168, 190–91, 190f; *vs.* national patterns, 167–68, 170–71, 171t, 187, 188, 190–91, 191f, 193–95, 193f, 224; neighborhood-specific effect, 167, 174–76, 175t, 177; Puerto Ricans, 19, 43–44, 43t, 69, 84t, 85, 170, 171t, 224; race effect on, 167; among women, 21, 23, 44–46, 45t, 84t, 85, 167, 169, 171–73, 171t, 176–77, 190f, 191, 193–95, 193f (*see also* women)
labor market reentry rates: ethno-racial group differentials, 204–5, 205f; *vs.* national patterns, 202–3, 202f, 214; among women, 202f, 203–4
labor market standing: of blacks, 168, 180 (*see also* blacks, discrimination toward); correlation with education investments, 19, 181–83; correlation with employment experience, 4, 19–20, 191–93, 192f; between minorities, 180, 212, 213–14; of minorities *vs.* whites, 19, 21, 180, 181, 183, 188–89
Laub, John H., 9
Lein, Laura, 161, 163, 199
Lewis, Oscar, 14
life course perspective: accumulated disadvantages in, 27, 167–68, 213–14, 221–22; discrimination in, 168, 183 (*see also* job discrimination); labor market outcomes in, 183; pathways, construct of (Hogan and Astone), 9; poverty effects *vs.* race effects, 120, 220, 222; sequencing of family formation in, 10, 23; "turning points," construct of (Sampson and Laub), 9
life table methods, 152, 196
Losing Ground (Murray), 14

Making Ends Meet (Edin and Lein), 163, 199
male marriageable pool hypothesis (Wilson and Neckerman), 6, 127
manufacturing jobs, decline in, 48–52, 50f, 51f
marriage pathway to family life, 13, 109–11, 110f, 111f; age-specific effects on, 109; among ethno-racial groups, 3, 84t, 94, 110–11, 114f, 115, 146; gender effects on, 110; group effects of, 110–11; *vs.* national patterns, 3, 110, 110f, 111f, 114f; place effect on, 110, 114, 119, 119f; probability of, 115–17, 116f; and probability of future welfare use, 146–47, 146f; related household income, 121, 121f; significance of timing of, 10, 15, 109–11; and welfare participation, 15, 126, 146–47, 146f, 155–56, 155f
Massey, Douglas, 8, 22n.13, 31, 31n.4, 177, 177n.6, 180n.8
McLanahan, Sara S., 10–11, 101
Mexicans: educational levels of, 19, 41t, 42, 84t, 85–87, 102, 103t, 104f, 105, 224; family formation patterns among, 112–13, 115–17, 116f, 146; family structure of, 46, 97, 146; immigration to Chicago, 32–33; labor force participation of, 19, 22–23, 43–44, 84t, 85, 170–72, 171t, 193–95, 236; pathways to family life, 102, 110–13, 110f, 111f; poverty among, 47, 95–96, 96f, 236–37; significance of immigrant status, 86; wages of, 23, 208–9, 208t, 211, 225, 236; welfare use among, 84t, 85, 86, 149–50, 153–54; and working-class neighborhoods, 35–40, 38f, 39f, 67, 70
minimum-variance algorithm (Ward), 55n.9

minority-group status, delimited opportunities for, 4, 30, 44, 47, 164–65, 167, 168, 211, 212, 215n.23, 216, 221
mismatch hypothesis (Holzer), 48
Multi-City Study of Urban Equality, 232, 234, 234n.6, 237

National Opinion Research Center, 72n.2
National Survey of Families and Households (NSFH), 71–72, 79–83, 82t; compared with UPFLS, 79–80, 81–83
Neckerman, Kathryn M., 127
neighborhoods: degrees of poverty among, 26, 54, 100–101, 137–38, 137f, 141; distribution of groups within, 35–40, 84t, 85; downgrading of, 63–65, 66t, 67, 70; labor force participation in, 174–76, 175t; pathways to change of, 63–69, 66t, 70; race-specific effects, 63–69, 70; social ecology of, 30, 55–63, 64t; spatial distribution of, 60f, 61f, 62f, 137; spatial polarization of, 30, 63–65, 64t, 68–69, 138, 222 (*see also* segregation, residential); typology of, 54, 56–59, 57t; upgrading of, 62–63, 65–68, 66t, 70
neighborhood typology: gentrifying yuppie, 58, 59, 63, 63f, 65, 68, 84t, 85; ghetto underclass, 59, 60–63, 66, 84t, 85 (*see also* ghetto underclass neighborhoods); middle-class, 56–58, 59, 63; working-class, 58, 59–60, 61f, 62, 63

opportunity, economic: ethno-racial inequalities in, 29; factors influencing, 194, 222; minority status and, 44, 47; race as delimiting factor, 7, 167, 211–12, 219, 221, 234; residential segregation and, 8, 40, 163, 177, 219; skills differentials and, 43; spatial distribution of, 177, 195, 223, 234
opportunity, inner-city attitudes toward, 165–66, 178–80, 179t, 225–32, 227t, 231f (*see also* SOS respondents)
out-of-wedlock births. *See* birth pathway to family life; teenage childbirth

parental supervision, 99–101, 100t
pathways, 8–9, 129, 221. *See also* life course perspective
pathways to family life. *See* birth pathway to family life; marriage pathway to family life

Personal Responsibility and Work Opportunity Reconciliation Act of 1996 (PRWORA), 14n.7, 133, 133n.1, 147n.2, 151
person-year files, 189
poverty: concentration of, 46–47, 53; correlation with nonmarital fertility, 101, 111–12; correlation with welfare participation, 45–46, 46t; effects of childhood exposure on, 3, 10, 221; effects of family structure on, 10, 11, 141, 221; explanations for, 7, 11; group probability of, 236; intergenerational transmission of, 3, 7fn.3, 9, 92–93, 120; *vs.* national patterns, 2, 25–26, 72, 78–79, 80–81, 81n.9; spread of, 68, 235–37. *See also* intergenerational transmission of poverty
poverty, childhood exposure to: correlation with alternative pathways to family life, 93, 104, 114, 118–19, 119f; and educational attainment, 93–94, 107–8, 107f, 184; effects on, 3, 10; ethno-racial differences in, 95, 96t, 220, 222; *vs.* national patterns, 93, 95–96, 96t, 98, 118–19, 119f, 136–37, 142–43, 143f; as vehicle of poverty, 93–94, 108, 142–43, 143f; and welfare use, 142–43, 142f, 235
poverty rate, 46–47, 46t, 235–36
predicted wages. *See* wages, predicted
public assistance. *See* welfare participation
public housing, 177, 180n.8
Puerto Ricans: differences with Mexican experience, 33, 173–74; discrimination toward, 167, 211; educational levels of, 19, 41t, 42, 84t, 85, 102, 103t, 224; family formation patterns among, 112, 112f, 113f, 115–17, 116f, 146; immigration to Chicago, 32–33; labor force participation of, 19, 43–44, 43t, 69, 84t, 85, 170, 171–73, 171t, 180, 193–95, 224; poverty among, 47; welfare use among, 46t, 47, 84t, 85, 131, 149, 153–55, 163

queuing theory, 22

Rainwater, Lee, 13n.6, 17–18
Rank, Mark, 15n.8
Rein, Martin, 13n.6, 17–18
"reservation wages," 207–9, 208t, 211–12, 216

Sampson, Robert J., 9
Sandburg, Carl, 28

Sandefur, Gary, 10–11, 101
"second ghetto," 35
segregation, residential, 31, 223–23; and concentrated urban poverty, 22n.13, 148–49, 163, 177–78, 219, 222; effect of on economic opportunity, 40, 163, 222. *See also* blacks, residential segregation of
siblings, influence of, 97
single-parent families: and birth pathway among next generation, 117, 118f; correlation with high school noncompletion, 101; among ethno-racial groups, 3, 95; group probability of, 93; *vs.* national patterns, 93, 95; and risk of welfare use, 141–42, 142f
skills mismatch, 42, 181, 224
"social buffer" hypothesis (Wilson), 9
social ecology, changes in, 67–68, 70
socialization theory, 156
Social Opportunity Survey (SOS), 71, 73, 74–76, 75f, 163–64, 226–32. *See also* opportunity, attitudes toward; SOS respondents
SOS respondents: Alyson, 215; Belinda, 228; Bernardino, 214; Billy, 120; Brianna, 166, 168, 228–29; Carol, 177; Cassandra, 130; Felicia, 92; Gina, 1; Gloria, 165–66; Ilana, 158–59; Irene, 161; Jane, 151; John, 228; Lashandra, 206; Laura, 177; Letoya, 1; Manuel, 1, 226; Marcelina, 216; Mary, 130; Ramiro, 205, 214; Rebecca, 108, 109; Rubén, 226; Sabrina, 2; Shala, 127, 163; Susan, 180–81, 230; Toribio, 230–31; Wendy, 162
South Side (Chicago), 35, 60–63
spatial distribution of jobs. *See* jobs
spatial mismatch hypothesis, 195, 224
spatial polarization of neighborhoods, 30, 53, 68, 70, 150
Standard Metropolitan Statistical Area (SMSA), 29, 35
structure, role of in poverty, 5, 6–8, 11, 11n4, 22. *See also* queuing theory; urban underclass debate
suburbanization, 33, 34
suburbs: job growth in, 48–53, 49t, 177–78 (*see also* Chicago, job decline in *vs.* in suburbs); minorities in, 34t, 40, 86
Supplemental Security Income, 13
survivor functions, 152

"taste for welfare," 127, 131, 141, 149–50, 156
teenage childbirth: and childhood exposure to poverty, 119, 119f; effect on educational attainment, 94, 101; effects on labor market options, 23, 102; among ethno-racial groups, 102–4, 103t; influence of family of orientation on, 94, 101, 104, 117–18, 118f; as mechanism of intergenerational poverty, 101; *vs.* national patterns, 102–4, 103t, 114; as pathway to welfare use, 102, 235; probability of, 115–17, 116f
Temporary Assistance for Needy Families (TANF), 14n.7, 133n.1, 151, 234–35
training, general *vs.* specific, 20–21, 187
"turning points" (Sampson and Laub), 9, 24, 27

Underground Railroad, and black migration to Chicago, 30
Unemployment Insurance (UI), 48
unskilled labor, demand for, 69, 181
Urban Poverty and Family Life Study, 71, 71n.1
Urban Poverty and Family Life Survey (UPFLS), 25, 26, 30, 71, 72–74, 76–79, 81–85, 82t, 84t; parents as focus in, 78; sampling framework of, 72–73, 76–77, 77n.7, 82t; variation of neighborhoods in, 76–77, 90
urban residents, poor (NSFH subsample): educational levels among, 87, 89t, 102, 103t, 107–8, 107f; family formation patterns of, 94, 103t; labor participation of, 89, 90; marital status among, 87–89, 89t, 94; minority disadvantages, 89, 90; welfare use among, 89t, 90
urban residents (NSFH respondents): educational levels, 86, 88t, 102, 103t, 107–8, 107f, 220t; family formation patterns of, 86, 88t, 94, 103t, 220t; labor force participation among, 86, 88t, 220t; welfare use among, 86, 87, 88t, 90, 220t
urban underclass, *vs.* urban poor, 130
urban underclass debate, 2, 5–9, 134

wage expectations, 168–69, 183, 208–9, 211
wage inequality, 4, 22, 31, 169, 208–9, 208t, 225
wages, predicted, 210–11, 210f, 211n.22
"waiting times," 200
waiting times, structural *vs.* individual reasons for, 200–201, 201f
welfare, intergenerational transfer of, 15–17;

as effect of early exposure to poverty, 18, 132, 142–43, 143f; *vs.* national patterns, 143
"welfare class" (Rein and Rainwater), 13, 13n.6
welfare participation: behavioral effects of, 16, 16n.9; among blacks, 46t, 47, 69, 84t, 85, 96, 126–27, 126f, 131–32, 134t, 135–41, 149, 153–55, 163, 220t, 222–23; chronicity among ethno-racial groups, 153–54, 154f, 163; duration of spells, 15, 16, 133, 151, 152–57, 152–55, 152f, 154f, 155f, 223; effect of disadvantaged background on, 131, 139, 142–43, 143f, 162–63, 220, 222, 223; effect of divorce on, 145; effect of limited employment opportunities on, 163–64, 220–21, 222; effect of pathways to family life, 155–54, 155f; effect of segregation on, 131, 138, 163; ethno-racial dispositions as explanation for, 11; exits from, 15, 146, 148, 153; explanations for, 136, 144, 156–57; family structure effect on, 123, 124f, 141–42, 142f; gender differences in, 123–24, 131, 135, 139, 140f, 141–42, 142f, 149; and growth of urban underclass, 17; inner-city attitudes toward, 157–62, 158t, 159t (*see also* SOS respondents); intergenerational continuation of, 17–18, 101, 131, 156–57; and joblessness, 147–48, 149 (*see also* "concentration effects"); in life course perspective, 4, 13–18, 148, 162–63; and marital status, 123, 124f; among Mexicans, 131, 134t, 135–36, 138, 139; *vs.* national patterns, 3, 4, 81–82, 131, 134t, 135, 136, 139, 141, 142f, 143f, 148–50, 223; and neighborhood poverty rate, 137–38, 137t, 141; and pathways to family life, 155–57, 155f, 223; patterns of, 15–16, 132–33, 152–53, 152f; pervasiveness of, 134–36, 134t; poverty effect on, 148–49; preference for, 127, 138–39, 149–50, 156, 158–62, 225; prevalence of in ghetto neighborhoods, 137t, 141, 148–49; probability of, according to life course perspective, 10; among Puerto Ricans, 69, 131, 134t, 135–36, 138, 140, 140f; rates of, 134–36, 134t, 137t; among whites, 131, 134t, 135–36, 138, 139. *See also* Aid to Families with Dependent Children (AFDC)
West Side (Chicago), 35, 60–63
When Work Disappears (Wilson), 7
whites: educational levels among, 84t, 85, 103t, 224; family formation patterns, 115–17, 116f; labor force participation among, 19, 23, 43–44, 43t, 84t, 85, 171t; poverty among, 47, 96; welfare participation of, 97
willingness to work, 7, 132, 207–8 (*see also* "reservation wages")
Wilson, W. J., 5–6, 25, 65, 157, 215, 223; "concentration effects," 6, 106, 149; and distinguishing features of inner-city poor, 5, 7, 12, 26; male marriageable pool hypothesis, 10, 127; "social buffer" hypothesis, 9, 107; on spatial concentration of jobs and people, 157, 173, 177, 223; on structural factors perpetuating urban poverty, 5–6, 22, 180n.9, 202–3, 218–19; and Urban Poverty and Family Life Study, 71, 180n.9
women: group labor force participation of, 23, 44–46, 84t, 85, 176–77; labor force participation, *vs.* national patterns, 171–72, 171t, 190f, 191; labor force participation among (*see* labor force participation, among women); pathways to family life (*see* birth pathway to family life; marriage pathway to family life); wage inequality among, 208–9, 208t; welfare use among (*see* Aid to Families with Dependent Children (AFDC); welfare participation)
work exit rates: ethno-racial differentials in, 196–98, 197f, 199f; gender differentials in, 198–99, 201, 201f; group differentials among, 196–98, 197f, 199f; *vs.* national patterns, 197–99, 199f. *See also* labor market instability; labor market outcomes, differences in; waiting times
work experience. *See* human capital investments, and differential work experience

Zinn, Baca, 11n.4